Monitoring and Modelling Lakes and Coastal Environments

Preface

Lakes and coastal environment play a vital role in the global ecosystems. Their importance has been recognized in the maintenance of biodiversity, ecology, hydrology and recreation. They provide habitat for wide variety of flora and fauna and help maintain the life cycle of many species. Lakes and coastal environment all over the world are ideal places for human habitation, fisheries, industries, shipping and recreation. Habitat environment of lakes and coastal environment are deteriorating due to their exploitative use and improper management. As rapid development and population growth continue in coastal areas, environmental degradation and over-exploitation will further erode the biodiversity and undermine the productivity of these unique ecosystems.

Lakes and coastal environments are transitional areas between dry terrestrial and permanent aquatic ecosystems and are recognised as highly productive. Their importance in socio-economic frontiers has been increasingly felt. These have been used as main source of water supply, food, fodder, fuel, fishery, aquaculture, timber production, transport, ecotourism, culture and heritage, research and educational values. The patterns of human occupancy and activity in the relatively narrow coastal strip have significantly affected the coastal environment.

Due to increased use of lakes and coastal environment and exploitation of their resources for various economic growths, these ecosystems are under severe stress. The stress may further increase in coming years because of urbanization, industrial growth, transportation, agriculture, housing etc. Unless timely corrective measures are taken, over-exploitation and environmental degradation will erode these ecosystems, which in turn will affect their productivity.

Various efforts are currently underway to develop technologies and systems for successful management of lakes and coastal environments—both at national and international levels. However, conflicting interests in the use of their resources have led to further worsening of the problems facing lakes and coastal environments. The important issues from a coastal environment perspective are livestock raising and agriculture in the coastal zone; the planning, control, and servicing of urban development in this area; the planning and assessment of major coastal facilities such as industrial projects, tourist facilities, and ports; the development of fisheries in the coastal zone; and the conservation of coastal and near-shore natural resources. Therefore, it is very important to involve all of those concerned in the process of restoration, conservation and management of lakes and coastal environments. Thus, there is an imminent need to promote regional linkages, develop strategic partnership and follow good practices in the conservation and managements of lakes and coastal environments. It is also essential to establish new, and strengthen ongoing regional and international co-operation, linkages and strategic partnership between governments, international agencies, universities, research institutions, non-governmental organizations, private sectors, local communities, and individuals.

In view of the interdisciplinary approaches in the study of lakes and coastal environment and the rapid progress taking place in the concerned branches of science and engineering, it is hard for researchers to keep abreast with all the developments without periodic interactions and discussion. With the view to promote such an interaction, the Lake series conference, a biennial event, was started by the Karnataka Environment Research Foundation and the Centre for Ecological Sciences, Indian Institute of Science, Bangalore in the year 2000. The main objective was to provide a forum to present and share scientific knowledge in aquatic ecosystem conservation, restoration and management. Lake 2002, the second conference was organized by the Centre for Ecological Sciences, Indian Institute of Science, Bangalore. Lake 2004 is the third in the series. LAKE-2004, an international conference on "Conservation, Restoration and Management of Lakes and Coastal Wetlands", was organised in Bhubaneswar, Orissa during the period 9-13 December, 2004. Distinguished scientists, managers, social workers and administrators from across the world participated in the deliberations and discussions on different scientific and socio-economic aspects of lakes and coastal wetlands. Eminent scientists who have made significant contributions in their respective fields presented ten keynote lectures. A special session on Chilika Lake was arranged with two invited lectures. Besides, there were 65 oral presentations and 28 poster presentations on a wide range of topics such as biodiversity, coastal engineering, limnology, monitoring and modelling, remote sensing and geographical information system, water quality, watershed hydrology and hydrogeology, environmental protection laws and policy options, socio-economic considerations, people's participation and awareness, recreation and ecotourism and management aspects. This book is mainly an outcome of the conference LAKE-2004. The papers presented in the conference were peer reviewed by the editorial board and 18 papers mostly on monitoring and modelling aspects of lakes and coastal environment were included for publication in a book form.

I hope the book fulfils a gap in our current understanding and knowledge on monitoring and modelling lakes and coastal environments for their conservation, restoration and management. The book is intended as an appeal to all scientists, managers and social workers to entertain a more global and holistic perspective and to adopt a macroscopic outlook on their approach to conservation, restoration and management of lakes and coastal environment.

I wish to record my sincere thanks to the organisers, co-organisers, sponsors and the scientific community for their active participation in the deliberation and discussion and for contributing papers to LAKE-2004 and to this book. The conference was organised by the Institute of Mathematics and Applications, Bhubaneswar. It was organised in co-operation with the Berhampur University, Berhampur; Chilika Development Authority, Bhubaneswar; Orissa Remote Sensing Application Centre, Bhubaneswar; Karnataka Environment Research Foundation, Bangalore; Centre for Ecological Sciences, Indian Institute of Science, Bangalore; and Centre for Atmospheric Sciences, IIT, Delhi. The organizing committee express their appreciation for the support afforded by the sponsors, the Commonwealth of Learning, Canada; Ministry of Environment and Forest, Government of India; Department of Science and Technology, Government of India; Department of Biotechnology, Government of India; Indian Space Research Organisation; Council of Scientific and Industrial Research, Government of India; Department of Ocean Development, Government of India; and the Indian Institute of Technology, Kharagpur. I am indebted to the members of the editorial board for their conscientious effort in reviewing the manuscripts, valuable suggestions and overall guidance in editing the book without which the publication would not have been possible.

Pratap K. Mohanty
Department of Marine Sciences
Berhampur University, Berhampur-760007
Orissa, India

3. YIELD SCENARIO

Data collected by the Department of Fisheries, Orissa indicated that the highest catch (8926 MT) during 1986-87 declined to 1274 MT during 1995-96, which was recorded as the lowest catch in the past. Jhingran and Natarajan (1969) suggested to separate prawn component from the fish component while collecting catch statistics in order to study the fishery trend judiciously. In the present study, prawns, shrimps and crabs were monitored as a separate 'shellfish' component and fish component (sharks, batoid fishes and bony fishes) were separately dealt with.

3.1 Prior to the Opening of the New Lake Mouth

Fish and shellfish (prawns, shrimps and crabs) landings in Chilika lake (1996-97 to 1999-00) prior to opening of the new lake mouth in September 2000 and after opening of the new lake mouth (2000-01 to 2003-04) were analyzed to study the impact of hydrological intervention on the fisheries of the lake. Pre-mouth data (averaged for four years) as presented in Table 1, shows drastically low fish and shellfish catch of 1489.07 MT and 197.29 MT respectively, which are less by 79.55% and 87.99% respectively from the fish and shellfish landings in 1986-87. This indicated that the prawn and crab fishery were more affected than fish during the period from 1987-88 to 1999-2000. Shellfish yield formed 11.70% in the total fisheries output during 1996-97 to 1999-2000 (Table 1). Cluepeoids, catfishes and mullets dominated the catch with 23.87%, 11.80% and 10.16% respectively. Freshwater fishes such as murrels, featherbacks and miscellaneous forage fishes together constituted 19.61% which was considerably higher. More freshwater condition and freshwater weed infested environment in the northern sector encouraged the population growth of murrels, featherbacks and weed fishes. One invasive freshwater species (*Oriochromis mosambica*) continued to propagate in the central sector and northern sector during the pre-mouth period under low salinity condition. Catch analysis for four years before new mouth indicated that the abundance of three fish groups (clupeoids, mullets and catfishes) were relatively stable despite faster degradation of the ecosystem. Mohanty et al. (2003), while evaluating commercial fish landings from Chilika lake (1997-98 to 1999-2000), reported that the northern sector, central sector, southern sector and outer channel sector shared 32%, 45%, 14% and 9% of the total catch respectively. Since penaeid prawns and portunid crabs breed in the sea and their juveniles are recruited from the sea into the lake, the prawn and crab population were decreased during this period due to recruitment failure as the lake mouth was shifted very far (about 30 km) from the lake proper and the confluence point of outer channel (recruitment route) at Magarmukh was silt-choked. The overall estimated productivity (fisheries output) was only 1.83 MT sq km^{-1} before restoration.

Table 1: Relative catch composition (by weight) of fish and shellfish in Chilika lake before and after opening of the new lake mouth (hydrological intervention)

Fish and shellfish of commercial importance (Group/species)	*Four years average catch (MT) before new mouth*		*Four years average catch (MT) after new mouth*		*% increase in Catch*	*% increase/ decrease in composition*
	Catch (MT)	*% Composition*	*Catch (MT)*	*% Composition*		
Fish						
Mullets	151.24	10.16	761.98	9.62	403.82	-5.31
Clupeoids	355.39	23.87	2254.13	28.47	534.27	19.27
Perches	132.89	8.92	458.24	5.79	244.83	-35.09
Threadfins (Polynemids)	67.65	4.54	336.62	4.25	397.59	-6.39
Croakers (Sciaenids)	101.64	6.83	741.18	9.36	629.22	37.04
Beloniformes (Needle fishes and Half beaks)	66.83	4.49	354.55	4.48	430.53	-0.22
Catfishes	175.74	11.80	1453.89	18.35	727.30	55.51
Tripod fish (*Triacanthes* Sp.)	46.98	3.15	402.49	5.08	756.73	61.27
Cichlids	98.81	6.63	296.05	3.74	199.61	-43.59
Murrels	55.49	3.73	187.39	2.37	237.70	-36.46
Feather backs (*Notopterus* sp.)	86.79	5.83	295.07	3.73	239.98	-36.02
Others	149.62	10.05	377.03	4.76	151.99	-52.64
Total fish landing	1489.07	100.00	7918.62	100.00	431.78	
Shellfish						
Penaeus monodon (Giant tiger prawn)	20.56	10.43	288.02	11.25	1300.87	7.86
Penaeus (Fenneropenaeus) Indicus (Indian white Shrimp)	27.39	13.86	421.25	16.45	1437.97	18.69
Metapenaeus monoceros (Brown Shrimp)	67.66	34.30	802.31	31.32	1085.80	-8.69
Metapenaeus dobsoni (Soft brown shrimp)	71.40	36.20	735.88	28.74	930.64	-20.61
Non-penaeid prawns (*Macrobrachium* sp.)	NA	NA	186.10	7.27	—	—
Mudcrabs (*Scylla* sp.)	10.28	5.21	127.48	4.97	1140.08	-4.61
Total shellfish landing	197.29	100.00	2561.04	100.00	1198.11	

NA – not available

3.2 After Opening of the New Lake Mouth

Post-new mouth monitoring of fisheries in the Chilika lake indicated that the hydrological intervention resulted in a far earlier recruitment of juvenile prawns and crabs into Chilika lake, and their retention for a longer period, presumably due to the maintenance of higher salinity and nutrient status for longer periods. Average catch for four years after opening of the new mouth (Table 1) worked out to 10,479.66 MT indicating 521.44% increase over pre-mouth data. Fish and shellfish catch during post restoration phase indicated spectacular leap with 431.78% and 1198.11% increase in comparison to the pre-restoration catch. Average yield as worked out from the average fisheries output (productivity) during post restoration period (2000-01 to 2003-04) was 11.35 MT sqkm^{-1} which indicated 520.22% increase over pre-mouth data. Sector-wise catches indicated that the northern sector, central sector, southern sector and the outer channel sector contributed 41.84%, 47.13%, 8.86% and 2.17% respectively to the total catch. Compared to the pre-restoration situation, catches from northern sector and central sector were significantly increased during post-restoration period, which can be attributed to elevation in salinity gradient in these sectors and clearance of freshwater weeds from the northern sector. Fish populations, particularly those belonging to brackishwater habitat presumably were distributed more evenly in both central and northern sectors due to improvement in hydro-biological conditions after opening of the new mouth.

4. FISHERY BIODIVERSITY STATUS

The unique ecological complex, existence of four ecological sectors, adequate availability of natural food elements, openness of the lake to two hydrological systems (marine and freshwater) resulting in two antagonistic hydrological processes, with penetration of fish and shellfish faunas respectively from marine and inland origin and cyclical change of salinity gradient, provide diverse habitat conditions in Chilika lake for both migratory and resident/ endemic fish and shellfish faunas with greater diversity.

The ZSI (1914-24), CIFRI (1957-65), other individual workers (1954-86), Chilika Expedition by ZSI (1985-87) and CDA (1998-2000) documented 225 fish species, 24 prawns and shrimps and 28 crab species as occurring in the Chilika lake before opening of the new mouth. Migration and movement of fish and shellfish faunas from the sea to lake and vice-versa, permanent resident species of brackishwater habitat within the lake, movement between brackishwater and freshwater mainly for feeding and breeding purposes exhibit diversities in habitats and nature of occurrence. Thus fishery biodiversity in Chilika lake needs to be studied in terms of species, habitats and occurrences.

4.1 Pre-restoration Status

During 1914-24, ZSI carried out the pioneering work on faunal diversity of Chilika lake and documented 112 fish species, 24 prawn and shrimps and 26 crab species (Chaudhuri, 1916a, 1916b, 1916c, 1917, 1923; Hora, 1923 and Kemp, 1915). During the first fisheries investigation in Chilika lake by CIFRI (1957-65) and by some individual workers during 1954-86, 101 fish species were documented from Chilika lake (Koumans, 1941; Jones and Sujansinghani, 1945; Mitra, 1946; Devasundaram, 1954; Roy and Sahoo, 1957; Menon, 1961, Mishra, 1962, 1969, 1976a, 1976b; Jhingran and Natarajan, 1966, 1969; Rajan et al., 1968; Mohanty, 1973; Talwar and Kacker, 1984; Talwar and Jhingran, 1991; Ramarao, 1995; Reddy, 1995; Maya Deb, 1995 and Bhatta et al., 2001). Thus, the pre-restoration status of species diversity in fish and shellfish faunas in Chilika lake stood at 225 fish species under 149 genera, 72 families and 16 orders, 24 prawn and shrimp species comprising 13 genera, nine families and two suborder and 28 crab species distributed under 22 genera, nine families and one suborder (Table 2). Fish species included four sharks, eight batoid fishes and 213 bony fishes. Twenty four recorded prawn and shrimp species included five number penaeid prawns and 19 number non-penaeid prawns. Among 28 recorded crab species, food crabs belong to three families (Calippidae, Portunidae and Grapsidae).

Prior to eco-restoration of Chilika lake, the ecosystem was under severe threats, most of which had roots to the natural changes and human-induced activities. During this phase, the migration and recruitment routes, habitats were considerably affected along with decline in salinity regime and proliferation of freshwater weeds in the northern sector. River mouths and Magarmukh were heavily silted to adversely affect the normal functioning of the ecosystem. Such conditions are likely to result in significant changes in faunal diversity and habitats. However, no attempt was made in the past to carry out inventorisation survey of fish and shellfish faunas in Chilika lake, which was most needed during eco-degradation phase. Further information on fishery biodiversity with reference to habitat and occurrence are not available except some limited account published by Biswas (1995) and Mohanty (2002). After reviewing literatures on icthyo-faunal records of Chilika lake, Khora (2002) reported that a good number of recorded species is overlooked, many are synonymied and some are invalid. Hence, a thorough inventorisation and review of literatures on the documented species of fish, prawn and crab before restoration of Chilika lake was considered imperative.

During 1985-87, while carrying out survey under Chilika Expedition Project, ZSI collected 63 recorded fishes, 13 prawns and 11 crab species and added four new records of fishes and two new records of crab species. Later, CDA carried out organized inventorisation survey and reported eight new records of fish species before opening of the new mouth (Bhatta et al., 2001). Thus, the inventorisation survey (Table 2) carried out before restoration

Table 2: Biodiversity status (Number of species) of fish and shellfish in Chilika lake during pre- and post-restoration (hydrological intervention) phases

Status parameter	*Pre-restoration survey (1914-2000)*	*Post-restoration survey (2000-01 to 2003-04)*	*Total of both phases*	*% increase*
1. Recorded species	By ZSI, CIFRI & individual workers	By CDA (New record)		
Fish	225 (G/149, F/72, O/16)	43 (G/38, F/31, O/14)	268 (G/168, F/82, O/20)	19.11
Shrimp & Prawn	24 (G/13, F/9, SO/2)	4 (G/3, F/2, SO/2)	28 (G/14, F/9, SO/2)	16.66
Lobster	Not recorded	2 (G/1, F/1, SO/1)	2 (G/1, F/1, SO/1)	00.00
Crab	28 (G/22, F/9, SO/1)	7 (G/5, F/3, SO/1)	35 (G/23, F/29, SO/1)	25.00
2. Inventorisation of recorded species	Last survey by ZSI (1985-87)	CDA survey (2000-01 to 2003-04)	Same as inventorisation CDA survey by	
Fish	71 (G/60, F/43, O/13)	187 (G/110, F/71, O/20)	187 (69.78%)*	163.38
Shrimp & Prawn	13 (G/6, F/5, SO/2)	18 (G/8, F/5, SO/2)	18 (64.28%)*	38.46
Lobster	No record	2 (G/1, F/1, SO/1) (First time record)	2 (100.00%)*	00.00
Crab	11 (G/11, F/6, SO/1)	14 (G/9, F/5, SO/1)	14 (40.00%)*	27.27

G – Genus; F – Family; O – Order; SO – Suborder; ZSI – Zoological Survey of India; CIFRI – Central Inland Fisheries Research Institute; CDA – Chilika Development Authority; *Figures in parenthesis indicate percentage to total record under column 4

reported collection of 71 fish species under 60 genera, 43 families and 13 orders, 13 prawns and shrimps distributed under six genera, five families and two sub-orders, and 11 crab species belonging to 11 genera, six families and onė sub-order.

4.2 Post Restoration Status

The inventorisation survey for fish and shellfish faunas initiated by CDA during 1998 was continued in Chilika lake during the post restoration phase (Tables 3 and 4). In total, 187 fish species under 110 genera, 71 families and 20 orders, 18 prawn and shrimp species under eight genera, five families and two sub-orders and 14 crab species under nine genera, five families and one sub-order, totalling to 221 number of fish and shellfish species, were documented. The inventorisation survey during post-restoration phase indicated recovery of 69.78% fish, 64.28% prawns and shrimps, 100% lobsters and 40.00% crab species. For the first time, two species of Indian spiny lobsters (*Panulirus polyphagus* and *Panulirus ornatus*) under family Palinuridae and sub order Macrura were collected from the Chilika lake during 2003. A total of 56 new records of fish, prawn, shrimps, lobsters and crabs were documented along with collection of materials during 2000-01 to 2003-04. The new records (Table 4) include 43 number of fish species under 38 genera, 31 families and 14 orders, four number prawn and shrimps comprising three genera, two families and two sub orders, two lobster species under single genera, family and suborder and seven crab species belonging to five genera, three families and one sub order (Table 4). Inventorisation of already recorded species before restoration (Table 3) documented 144 fish species, 14 prawn and shrimp species and seven crab species totalling to 165 species of fish and shellfish along with collection of materials.

Biodiversity status with regard to habitat and occurrence of fish and shellfish faunas in Chilika lake as observed from the inventorisation survey carried out by ZSI during 1985-87 before restoration and survey conducted by CDA during 2000-01 to 2003-04 after restoration is documented in Table 5. Fish species belonging to marine-brackish water habitat continued to dominate both during pre- and post-restoration phases with 31.55% and 33.16% respectively. Similarly fishes belonging to brackishwater-marine and freshwater-brackish water habitats were stable in their composition during both phases. Relative abundance of freshwater species decreased from 14.67% during pre-restoration phase to 13.67% during post-restoration phase, while species moving from brackishwater to freshwater habitat were drastically reduced from 5.78% to 2.14%. Out of several penaeid shrimp species observed during post-restoration phase, five were commercial, contributing more than 94% to the total prawns and shrimps catch which constituted 20.84% in the prawn and shrimp faunal diversity. One freshwater prawn (*Macrobrachium malcolmasonii*) contributing to the commercial prawn catch constituted 8.33% of the faunal diversity. However, small sized prawns belonging to freshwater-

Table 3: Inventorisation of recorded fish and shellfish faunas of Chilika lake during post-restoration period

Family	*Species*	*H & O Status*
Fishes		
1 Carcharhinidae	1. *Scoliodon laticaudas* (Muller & Henle)	M,R
2 Dasyatididae	2. *Himantura uarnak* (Forsskal)	M,VR
	3. *Himantura walga* (Muller & Henle)	M,VR
3 Myliobatididae	4. *Aetobatus flagellum* (Bloch & Schneider)	M,VR
	5. *Aetomylaeus nichofii* (Bloch & Schneider)	M,R
4 Notopteridae	6. *Notopterus notopterus* (Pallas)*	F,A
	7. *Notopterus chitala* (Hamilton-Buchanan)	F,VR
5 Elopidae	8. *Elops machnata* (Forsskal)	BM,R
6 Megalopidae	9. *Megalops cyprinoides* (Broussonet)	BM,R
7 Anguillidae	10. *Anguilla bengalensis* (Gray)	MB,A
	11. *Anguilla bicolour bicolour* (Mc Clelland)	MB,R
8 Muraenidae	12. *Thyrsoidea macrura* (Bleeker)	M,R
9 Ophichthidae	13. *Pisodonophis boro* (Hamilton-Buchanan)	MB,R
10 Muraenesoscidae	14. *Muraenesox cinereus* (Forsskal)	MB,R
11 Clupeidae	15. *Anodontosoma chacunda* (Hamilton-Buchanan)*	MB,A
	16. *Corica soborna* (Hamilton-Buchanan)	B,R
	17. *Escualosa thoracata* (Valenciennes)	MB,R
	18. *Gonialosa manmina* (Hamilton-Buchanan)	MB,R
	19. *Gadusia chapra* (Hamilton-Buchanan)	F,A
	20. *Hilsa (Tenualosa)ilisha* (Hamilton-Buchanan)*	MB,A
	21. *Hilsa kelee* (cuvier)	MB,A
	22. *Nematalosa nasus* (Bloch)*	BM,A
12 Engraulidae	23. *Stolephorus bagenensis* Hardenberg	MB,A
	24. *Stolephorus commersonii* Lacepade	MB,A
	25. *Stolephorus dubiosus* Wongrantania*	MB,A
	26. *Stolephorus indicus* (Van Hasselt)	MB,R
	27. *Thryssa hamiltonii* (Gray)*	B,A
	28. *Thryssa mystax* (Schneider)	B,A
	29. *Thryssa polybranchialis* (Wongrantania)	MB,R
	30. *Thryssa purava* (Hamilton-Buchanan)	B,R
13 Chanidae	31. *Chanos chanos* (Forsskal)	BM,R
14 Cyprinidae	32. *Amblypharyngodon mola* (Hamilton)	F,R
	33. *Catla catla* (Hamilton-Buchanan)	F,R
	34. *Cirrhinus mrigala* (Hamilton-Buchanan)	F,R
	35. *Cirrhinus reba* (Hamilton-Buchanan)	F,R
	36. *Chela bacaila* (Hamilton-Buchanan)	F,R
	37. *Chela cachius* (Hamilton-Buchanan)	MB,R
	38. *Esomus danricus* (Hamilton-Buchanan)	B,R
	39. *Labeo rohita* (Hamilton-Buchanan)*	F,R
	40. *Labeo calbasu* (Hamilton-Buchanan)	FV,R
	41. *Puntius chola* (Hamilton-Buchanan)	FB,A
	42. *Puntius sarana* (Hamilton-Buchanan)	FB,R
	43. *Puntius sophore* (Hamilton-Buchanan)	FB,A
	44. *Puntius ticto* (Hamilton-Buchanan)	FB,A
	45. *Parluciosoma daniconius* (Hamilton-Buchanan)	B,R
	46. *Salmostoma bacaila* (Hamilton-Buchanan)	FB,R
15 Bagridae	47. *Aorichthys seenghala* (Sykes)*	F,R

(*contd.*)

Table 3 (*contd.*)

Family	*Species*	*H & O Status*
	48. *Mystus gulio* (Hamilton-Buchanan)*	BF,A
	49. *Mystus cavasius* (Hamilton-Buchanan)*	FB,R
16 Ariidae	50. *Mystus vittatus* (Bloch)	FB,R
	51. *Arius arius* (Hamilton-Buchanan)	MB,R
	52. *Arius tenuispinnis* Day	MB,R
	53. *Osteogeneniosus militaris* (Linnaeus)*	MB,A
17 Siluridae	54. *Ompok bimaculatus* (Bloch)	F,R
	55. *Ompok Pabda* (Hamilton)	F,R
	56. *Wallago attu* (Schneider)*	F,R
18 Schilbeidae	57. *Ailia coila* (Hamilton-Buchanan)	F,R
19 Pangasidae	58. *Pangasius pangasius* (Hamilton-Buchanan)	FB,A
20 Clariidae	59. *Clarias batrachus* (Linnaeus)	FB,R
21 Heteropneustidae	60. *Heteropneustes fossilis* (Bloch)	FB,R
22 Plotosidae	61. *Plotosus canius* (Hamilton-Buchanan)*	B,A
	62. *Plotosus lineatus* (Thunberg)*	B,A
23 Aplocheilidae	63. *Aplocheilus panchax* (Hamilton-Buchanan)	FB,A
24 Hemiramphidae	64. *Hyporhamphus limbatus* (Valenciennes)*	B,A
25 Belonidae	65. *Strongylura strongylura* (VanHasselt)*	B,A
	66. *Strongylura liura* (Blecker)*	FB,A
	67. *Xenentodon cancila* (Hamilton-Buchanan)	FB,A
26 Syngnathidae	68. *Hyppocampus brachyrhynchus* Duncker	MV,R
	69. *Ichthyocampus carce* (Hamilton-Buchanan)	BM,VR
27 Platycephalidae	70. *Platicephalus indicus* (Lineaeus)	MB,A
28 Centropomidae	71. *Lates calcarifer* (Bloch)*	MB,A
29 Ambassidae	72. *Ambassis commersoni cuvier*	MB,A
	73. *Ambassis gymnocephalus* (Lacepede)	MB,A
	74. *Chanda nama* (Hamilton-Buchanan)	MB,A
	75. *Pseudoambassi ranga* (Hamilton-Buchanan)	MB,A
30 Serranidae	76. *Epinephelus tauvina* (Forsskal)	M,R
31 Teraponidae	77. *Terapon jarbua* (Forsskal)*	MB,A
	78. *Terapon puta* (cuvier)*	MB,A
32 Sillaginidae	79. *Sillago sihama* (Forsskal)*	MB,A
33 Carangidae	80. *Carangoides paraeusteus* (Bennelt)	MB,A
	81. *Caranx carangus* (Bloch)	MB,A
	82. *Caranx sexfasciatus* (Quoy and Gairnard)	MB,R
	83. *Megalaspis cordyla* (Lineaeus)	B,R
	84. *Scomberoides tala* (Cuvier)	M,R
	85. *Selaroides leptolypis* (Cuvier)	M,R
34 Leiognathidae	86. *Leiognathus dussumieri* (Valenciennes)	M,A
	87. *Leiognathus equulus* (Forsskal)	M,A
35 Lutjanidae	88. *Lutjanus johni* (Bloch)*	MB,R
	89. *Lutjanus russelli* (Blecker)*	MB,R
	90. *Lutjanus argentimaculatus* (Forsskal)	MB,R
36 Datnioedidae	91. *Datnioides quadrifasciatus* (Sevastianov)*	B,A
37 Gerreidae	92. *Gerreomorpha setifer* (Hamilton-Buchanan)*	BM,A
	93. *Gerres oyena* (Forsskal)*	B,A
	94. *Gerres abbreviatus* (Blecker)	BM,R
	95. *Gerres filamentosus* (Cuvier)*	B,A
38 Haemulidae	96. *Pomadasys argenteus* (Forsskal)	MB,R

(*contd.*)

Table 3 (*contd.*)

39 Sparidae	97. *Acanthopagrus berda* (Forsskal)	MB,R
	98. *Crenidens crenidens* (Forsskal)*	MB,A
	99. *Rhabdosargus sarba* (forsskal)*	B,A.
40 Sciaenidae	100. *Daysciaena albida* (Cuvier)*	MB,A
	101. *Dendrophysa russeli* (Cuvier)*	BM,A
	102. *Paranibea semilactuosa* (Cuvier)	M,R
	103. *Protonibea diacanthus* (Lacepede)	M,R
41 Monodactylidae	104. *Monodactylus argenteus* (Linnaeus)	M,R
42 Drepanidae	105. *Drepane punctatus* (Linnaeus)	MB,R
43 Scatophagidae	106. *Scatophagus argus* (Linnaeus)	MB,R
44 Nandidae	107. *Nandus nandus* (Hamilton-Buchanan)	B,R
45 Cichlidae	108. *Oriochromis mossambicus* (Peters)*	FB,A
	109. *Etroplus suratensis* (Bloch)*	
46 Mugilidae	110. *Liza macrolepis* (Smith)*	BM,A
	111. *Liza melinoptera* (Valancienues)*	BM,A
	112. *Liza parsia* (Hamilton-Buchanan)*	BM,A
	113. *Liza subviridis* (Valenciennes)*	BM,A
	114. *Liza tade* (Forsskal)	MB,R
	115. *Mugil cephalus* (Linnaeus)*	BM,A
	116. *Rhinomugil corsula* (Hamilton-Buchanan)*	FB,A
	117. *Valamugil cunnesius* (Valenceinnes)*	BM,A
	118. *Valamugil speigleri* (Blecker)	MB,A
47 Scombridae	119. *Scomberomorus linolatus* (Cuvier)	M,R
48 Polynemidae	120. *Eleutheronema tetradactylum* (Shaw)*	M,A
	121. *Polydactylus indicus* (Shaw)	MB,A
49 Gobiidae	122. *Acentrogobius cyanomos* (Blecker)	B,R
	123. *Acentrogobius globiceps* (Hora)	B,R
	124. *Glossogobius giuris* (Hamilton-Buchanan)	FB,R
	125. *Olegolepis cylindriceps* (Hora)	BF,R
	126. *Oxyurichthys microlepis* (Blecker)	MB,VR
50 Trypauchenidae	127. *Tripauchen vagina* (Bloch and Schneider)	MB,VR
51 Siganidae	128. *Siganus javus* (Linnaeus)	MB,R
52 Anabantidae	129. *Anabas testudineus* (Bloch)	F,R
	130. *Anabas cobojius* (Hamilton-Buchanan)	F,VR
53 Belontidae	131. *Colisa fasciatus* (Schneider)	F,R
	132. *Colisa lalia* (Hamilton)	F,R
54 Channidae	133. *Channa striatus* (Bloch)*	FB,A
	134. *Channa punctatus* (Bloch)	FB,A
55 Mastacembelidae	135. *Macrognathus pancalus* (Hamilton-Buchanan)*	BF,A
	136. *Mastacembelus armatus* (lacepede)*	BF,R
56 Bothidae	137. *Pseudorhombus arius* (Hamilton-Buchanan)	MB,R
57 Cynoglossidae	138. *Cynoglossus puncticeps* (Richardson)	MB,A
58 Soleidae	139. *Euryglossa orientalis* (Bloch)	MB,R
59 Tricanthidae	140. *Triacanthus biaculeatus* (Bloch)*	B,A
60 Tetradontidae	141. *Chelonodon fluviatilis* (Hamilton-Buchanan)	MB,R
	142. *Chelonodon patoca* (Hamilton-Buchanan)	MB,R
	143. *Tetradon cutcutia* (Hamilton-Buchanan)	MB,R
	144. *Takifugu oblongus* (Bloch)	MB,R
Shrimps and Prawns		
1 Penaeidae	1. *Metapenaeus affinis* (H.Milne-Edwards)	MB,R

(*contd.*.)

Table 3 (*contd.*)

Family	*Species*	*H & O Status*
	2. *Metapenaeus dobsoni* (Micrs)*	BM,A
	3. *Metapenaeus monoceros* (Fabricius)*	BM,A
	4. *Penaeus(Fenneropenaeus) indicus*(H.Milne-Edwards)*	BM,A
	5. *Penaeus monodon* (Fabricius)*	BM,A
	6. *Penaeus Semisulcatus* (de-Haan)*	BM,R
2 Palaemonidae	7. *Macrobrachium lamarrei* (H.Milne-Edwards)*	B,A
	8. *Macrobrachium malcomsonii* (H.Milne-Edwards)*	F,R
	9. *Macrobrachium rude* (Heller)*	B,A
	10. *Exopalaemon styliferus* (H.Milne-Edwards)	FB,R
	11. *Periclimenes (Harpilius) demani* Kemp.	BB,R
3 Atyidae	12. *Caridina propinqua* de Man	BF,VR
4 Callianassidae	13. *Callianassa (Callichirus)maxima* H.Milne-Edwards	B,VR
5 Upogebiidae	14. *Upogebia (Upogebia) heterocheir* Kemp.	FB,VR
Crabs		
1 Calappidae	1. *Matuta planipes* Fabricus	MB,R
2 Leucosiidae	2. *Philyra alcocki* Kemp.	MB,VR
3 Ocypodidae	3. *Ocypoda macroara* (H.Milne-Edwards)	MB,R
4 Grapsidae	4. *Varuna litterate* (Fabricius)	MB,R
5 Portunidae	5. *Portunus pelagicus* (Linnaeus)*	MB,A
	6. *Scylla serrata* (forsskal)*	BM,A
	7. *Thalamita crenata* (Latre)	MB,A

* Commercial species; H&O: habitat and occurrence; M: marine; B: brackishwater; F: freshwater; MB: marine-brackishwater; BM: brackishwater-marine; FB: freshwater-brackishwater; BF: brackishwater-freshwater

brackishwater-freshwater habitat constituted 55.78% of the total prawn species before restoration. Shrimp species belonging to the Marine-brackishwater-marine habitat were increased during post-restoration phase forming 44.45%. One commercially important freshwater prawn (*Macrobrachium rosenbergii*) was found as a new record during post-restoration period. All crab species collected during pre- and post-new mouth phases belonged to marine-brackishwater-marine habitat. One species of mudcrab (*Scylla tranquebarica*), although was occurring in Chilika lake since the faunal diversity study by ZSI during 1914-24, it was hitherto not reported/documented as a separate mudcrab species, different from the commonly occurring species *Scylla serrata*. The species, *Scylla tranquebarica* (Fabr.) was earlier regarded as one of the four varieties of one mudcrab species, *Scylla serrata*. This controversy of species identification of the mudcrabs in the genus *Scylla* was ended when Fuseya and Watanabe (1996) and Fushimi and Watanabe (1999) confirmed by genetic variability studies in Japan that *Scylla tranquebarica* is distinctively a separate mudcrab species. Hence, this mudcrab species has been added as a new record to the crab faunas of Chilika lake during post restoration phases. Species diversity composition based on the status of occurrence (Table 5) indicated that abundantly occurring fish, prawn and

Table 4: New records of fishes, shrimps and prawns and crabs from Chilika lake after recent hydrological intervention during 2000-01 to 2003-04

Order	*Suborder*	*Family*	*Species*	*H & O Status*
Fish				
Carcharhiniformes (Ground sharks & allies)	Scyliorhinoidei	Sphyrnidae	*1. Sphyrna lewini* (Griffith and Smith)	M,VR
			2. *Sphyrnablochii* (Cuvier)	M,VR
Rajiformes (Skates & rays)		Rhinobatidae	3. *Rhynchobatus djeddensis* (Forsskal)	M,R
Myliobatiformes (Sting rays)		Dasyatididae	4. *Dasyatis marginatus* (Blyth)	M,R
Anguilliformes (Eels)	Anguilloidei	Muraenesocidae	5. *Muraenesox bagio* (Hamilton)	B,R
Clupeiformes (Herrings, Sardins, Shads & allies)	Clupeoidei	Clupeidae	6. *Sardinella fimbriatus* (Valenciennes)	M,VR
			7. *Sardinella longiceps* (Vol)	M,VR
			8. *Dussumieria elopsides* (Blecker)	B,R
			9. *Ehirava fluviatilis* Deraniyagala	MB,VR
			10. *Thryssa gautamiensis* (B. Rao)	MB,R
			11. *Thryssa setirostris* (Broussonet)	MB,A
Cypriniformes (Carps & minnows)	Cyprinoidei	Cyprinidae	12. *Labeo boga* (Hamilton)	F,R
			13. *Labeo gonius* (Hamilton)	F,R
			14. *Osteobrama cotio peninssularis* Silas.	F,VR
Siluriformes (Catfishes)		Sisoridae	15. *Bagarius yarellii* Sykes	F,R
Aulopiformes (Green eyes, Lizard fish and allies)		Synodontidae	16. *Trachinocephalus myops* (Forster)	M,VR
Atheriniformes (Silver sides)		Atherinidae	17. *Atherinomorous lacunosus* (Forster)	M,VR
			18. *Atherinomorus duodecimalis* (Valenciennes)	M,VR
Syngnathiformes (Pipe fishes)	Syngnathoidei	Syngnathidae	19. *Syngnathus cynospilus* Blecker	M,VR

(contd..)

Table 4 (*contd.*)

Order	*Suborder*	*Family*	*Species*	*H & O Status*
Synbranchiformes (Shore eels)		Synbranchidae	20. *Ophisternon bengalense* Mc Clelland	M,VR
Scorpaeniformes (Scorpion fishes & allies)	Scorpinoidei	Tetrarogidae	21. *Tetraroge niger* (Cuvier)	MB,R
			22. *Sugrundus rodri censis* (cuvier)	MB,VR
Perciformes (Perch-like fishes)	Percoidei	Serranidae	23. *Epinephelus coioides* (Hamilton)	M,R
		Sillaginidae	24. *Sillago vincenti* Mc. Kay	MB,VR
		carangidae	25. *Scomberoides commersonianus* (Lacepede)	M,VR
			26. *Scomberoides tol* (cuvier)	M,VR
			27. *Selar crumenophthalmus* (Bloch)	M,VR
			28. *Trachinotus mookalee* (Cuvier)	M,VR
		Leiognathi-dae	29. *Leiognathus bindus* (Valenciennes)	M,VR
		Gerreidae	30. *Gerres abbreviatus* (Blecker)	MB,VR
		Haemulidae	31. *Pomadasys kaakan* (Cuvier)	M,R
		Cichlidae	32. *Oriochromis mossambica* (Peters)	F,A
	Scombroidei	Trichuridae	33. *Eupleurogrammus glossodon* Blecker	M,VR
			34. *Lepturacanthus savala* (Cuvier)	M,VR
		Scombridae	35. *Rastrelliger kanagurta* (Cuvier)	M,VR
	Sphyraenoidei	Sphyraenidae	36. *Sphyraena jello* (Cuvier)	M,VR
	Polynemoidei	Polynemidae	37. *Polydactylus plebeius* (Broussonet)	MB,VR
	Gobioidei	Eleotridae	38. *Eleotris melanosoma* Blecker	M,VR
		Gobiidae	39. *Yongeichthys criniger* (Valenciennes)	B,VR
	Acanthuroidei	Acanthuridae	40. *Acanthurus mata* (Cuvier)	M,VR
		Siganidae	41. *Siganus canaliculatus* (Park)	M,R

(*contd..*)

Table 4 (*contd.*)

	Channoidei	Channidae	42. *Channa marulius* (Hamilton)	F,R
Pleuronectiformes (Left eye flounders)	Pleuronectodoidei	Bothidae	43. *Pseudorhombus triocellatus* (Bloch)	M,R
Shrimp/Prawn				
Decapoda	Penaeoidea	Penaeidae	1. *Penaeus (Melicertes) canaliculatus* (Oliver)	MB,R
			2. *Metapenaeus ensis* DeHaan	MB,R
	Caridea	Palaemonidae	3. *Macrobrachium rosenbergii* (DeMan)	F,R
			4. *Macrobrachium equidens* (Dana)	B,R
Lobster				
Decapoda	Macrura	Palinuridae	1. *Panulirus polyphagus* (Herbst)	M,VR
			2. *Panulirus ornatus* (Fabricius)	M,VR
Crab				
Decapoda	Brachyura	Portunidae	1. *Charybdis cruciata* (Herbst)	MB,R
			2. *Charybdis callianasa* (Herbst)	MB,R
			3. *Portunus sanguinolentus* (Herbst)	BM,A
			4. *Scylla tranquebarica* (Fabricius)	BM,A
			5. *Podophthalmus vigil* (Herbst)	MB,R
		Grapsidae	6. *Sesarma quadrata* (Fabricius)	MB
		Calappidae	7. *Mutata lunaris* (Forsskal)	MB

H&O: habitat and occurrence; M: marine; B: brackishwater; F: freshwater; MB: marine-brackishwater; BM: brackishwater-marine; FB: freshwater-brackishwater; BF: brackishwater-freshwater; R: rare; A: abundant; VR: very rare

Table 5: Biodiversity status (habitat and occurrence) of fish and shellfish in Chilika lake during pre- and post-restoration phases

Status parameter	*Pre-restoration (1914-2000)*			*Post-restoration (2000/01-2003/04)*			
	'Recorded species'			*'Inventorised species'*			
	Fish	*Shrimp & Prawn*	*Crab*	*Fish*	*Shrimp & Prawn*	*Lobster*	*Crab*
Percentage composition of species by habitat (%)							
Marine	21.33			21.93	5.55	100.00	
Brackishwater	9.78	8.33		11.76	33.33		
Freshwater	14.67	8.33		13.37	11.11		
Marine-Brackishwater	31.55	12.50	92.86	33.16	16.67		85.71
Brackishwater-Marine	5.78	20.84	7.14	5.88	27.78		14.29
Freshwater-Brackishwater	11.11	50.00		11.76	5.56		
Brackishwater-Freshwater	5.78			2.14			
Percentage composition of species by occurrence (%)							
Abundant	14.22	33.33	7.14	34.23	38.89		21.43
Rare	49.78	37.50	14.28	41.71	33.33		28.57
Very rare	36.00	29.17	78.58	24.06	27.78	100.00	50.00

Inventorisation of species during pre- and post-restoration phases are taken into consideration.

crab species increased during post-restoration phase. Abundantly occurring fish and crab species during pre-new mouth period increased significantly from 14.22 to 34.23% and from 7.14 to 21.43% respectively during post-new mouth period.

4.3 Species Richness

Species richness of estuaries and lagoons is defined as the number of species encountered at least once within ecosystem limits (Baran, 2000). Species richness (SR) in such aquatic ecosystems is dependant on the openness of the systems and characteristics of the spatio-temporal variation in salinity gradient. Chilika lake, which is estuarine in character, being influenced by three hydrological systems exhibits four distinctive ecological sectors. These four sectors show variations in species richness varying with the seasons. As observed from the species inventorisation survey undertaken during post-new mouth period, three sectors (northern, central and outer channel sectors) were more influenced by two antagonistic hydrological process resulting from freshwater inflow from rivers and catchment streams and sea water ingress from the sea. Freshwater inflow into the lake remains active and strong during June-November. Although low/feeble inflow of freshwater from rivers continues throughout the year, the sea water influx dominate during December-May. Therefore freshwater species in northern sector are gradually replaced by brackish water species from December onwards which are again gradually replaced by freshwater species coming in the river flows from July onwards. Similarly the outer channel sector is strongly influenced by both freshwater outflow and the sea water ingress during semi-diurnal flow tides. Therefore, the species richness of fish and shellfish faunas in those sectors showed wider variations, whereas least variation was observed in the southern sector due to weak freshwater inflow for shorter duration and restricted exchange of water through Palur canal (before renovation).

After opening of the new mouth near Ramabhartia, enhancement in species richness was observed, particularly in outer channel sector. Outer channel sector registered the highest species richness (62.44%) in summer and 54.75% in winter. Central sector came in the second order with 48.42-50.68% SR. Northern sector registered SR of 34% only during winter. Southern sector showed minimum variation (14.93-16.74%) in species richness. Higher species richness in the outer channel sector is due to entry of more marine species during summer and winter for feeding purposes, except few others for breeding (*Eleutheronema tetradactylum*, some *clupeoides* etc.). In general, the species richness is mostly due to a succession of species temporarily using these ecological sectors for feeding, spawning or shelter. Dominance of marine or freshwater species in the ecosystem depends on the strength of marine and freshwater.

5. POPULATION STRUCTURE

In estuarine ecosystem, the fish and shellfish population is structured according to a gradient of increasing or decreasing salinity (Baran, 2000). Analysis of commercial catches from Chilika lake during pre and post-restoration phases indicated the percentage compositions (by weight of catch) of different commercial fish groups (Table 1). The commercial fish catch is contributed by 12 fish groups namely, mullets, clupeoides, perches, threadfins, croakers, beloniformes, catfishes, tripodfish, cichlids, murrels, featherbacks and others. These 12 fish groups comprised 22 species before opening of new mouth, which were increased to 46 species during post new mouth period (Table 3) indicating 109.09% increase. Similarly, prawns and crabs showed 60.0% and 50.0% increase respectively during the post restoration period. Species of fish and shellfish contributing to the bulk of catches during post-restoration phase are presented in Table 3. Clupeoids continued to dominate the fish catch forming 23.87% and 28.47% during pre- and post-restoration phases respectively. Catfishes and mullets came in the second and third order respectively before and after the New Mouth. Clupeids in estuaries generally dominate the fish catch. Baran (2000) reported that in African estuaries clupeids are always numerically dominant and represent 61% to 85% of catches. But in Chilika lake, clupeids and engraulids dominated the commercial fish catch representing 24% to 28% by volume. Mullet, the most priced fish of Chilika lake has indicated slow recovery of its fishery during post-restoration period. The population enhancement of this fish depends on the successful spawning migration to the sea and recruitment of juveniles from the sea into the lake. Next to clupeoids, the population of sciaenids (croakers), catfishes and tripodfish significantly increased during post-restoration phase due to enhancement of environmental condition (habitat enhancement), successful migration, spawning, improvement in salinity gradient etc. Although seaward breeding migration of many economic fishes (*Mugil cephalus, Liza macrolepis,* smaller mullets, *Lates calcarifer, Eleutheronema tetrdactylum, Rhabdosargus, Sarba, Daysciaena albida* etc.) became effective during post-restoration phase, recovery of their fisheries have been delayed except *Eleutheroaema tetradactylum* and *Daysciarna albida*. Population of threadfins (*E. Tetradactylum*) and croakers (*Dendrophysa russelli* and *D. albida*) increased in the lake immediately after the new mouth since 2001-02. Relative catch (by weight) of croakers increased by 37.04% during post mouth phase, compared to pre-mouth years (Table 1). Although the population of mullets, perches and threadfins increased significantly during post-restoration years, their percentage composition in the total fish catch did not increase as the population of clupeoids, catfishes and tripodfish outwitted them. Resident species, freshwater species and migratory fishes dominated in sourthern, northern and in the central and outer channel sector respectively.

5.1 Recruitment Pattern of Some Economic Fishes

Jhingram and Natarajan (1969) reported the recruitment pattern of *Mugil cephalus, Liza macrolepis, E. tetradactylum, Hilsa ilisha, Nematalosa nasus, Mystus gulio, Lates calcarifer, Gerreomorpha setifer, Daysciaena albida* and *Etroplus suratensis.* The present study (catch analysis) indicated that the recruitment of *M. cephalus, L. macrolepis, E. tetradactylum, L. calcarifer* and *D. albida* was adversely affected during pre-restoration years due to several natural and human induced threats as described by Pattnaik (2000) and Mohanty et al. (2001). The opening of the new mouth and desiltation of outer channel at Maggarmukh in 2000 became favourable for effective recruitment of all economic fishes, except the cichlids (*E. suratersis*), whose population indicated gradual decline due to increase in salinity regime. Although *M. gulio* is a resident species, its population increased due to habitat improvement in northern sector after clearance of freshwater weeds and improvement in salinity. However, the recruitment pattern of migratory fishes did not indicate any shift from the earlier observations during the post restoration period and the recruitment of some important commercial fishes were observed as follows:

Recruitment of juveniles of *Mugil cephalus* took place from May onwards at modal size of 107 mm. A second batch of large juveniles with modal size of 195-220 mm were recruited from the sea during July-August when they contributed to the commercial catch in the outer channel and near Maggarmukh. *Liza macrolepis* marked their peak recruitment to the fishery during July-August at modal size of 176 mm. Higher fishing mortality of 180-250 mm size group might have affected the spawning in the subsequent years. *Eleutheronema tetradactylum* breeds both in the sea and in the lake performing sea-lake and vice versa migration. Recruitment took place twice in a year, first during March-May in the size range 45-152 mm and the other in August-October at 95-138 mm size range. The fish has a prolonged breeding habit. *Nematolosa nasus* constituted the highest percentage in the commercial catch (30-38%) which breed in the lake. The fish migrate into the lake from the sea in large number during February-May and form active fishery by recruitment. This species is not considered as a true resident one as the population is enhanced by immigrant batches from the sea. Relative abundance of this fish has indicated faster increase during post-new mouth phase. *Lates calcarifer* was observed during post-restoration period where two brood batches participated in the recruitment to the fishery; one during February-March and the other during July-August. Its fishery showed slow growth with distinct fluctuation. *Mystus gulio,* the common cat fish of Chilika lake, has prolonged breeding habit (June-December). The population dominated in the northern sector. Though the breeding grounds were located in the northern sector, Nalabana submerged island (bird sanctuary) was observed to be good nursery ground for *Mystus gulio* during early winter. The recruitments of *Daysciaena albida* was observed during post-restoration period, one in July-September (major recruitment) and the other in February indicating two spawning periods. Eco-restoration was found to have increased the population of this fish. Population of cichlid fish (*Etroplus suratensis*) continued to increase during pre-restoration years. But interestingly, the population rapidly decreased during post-restoration years, which could be attributed to increased salinity regime and decrease of submerged weeds from the western shore of the lake. Recruitment to the fishery was observed during May-June and during October-December.

5.2 Reappearance of 'Threatened Species'

In the process of continued ecological degradation of the Chilika lake during the last few decades before the hydrological intervention in 2000, six fish species namely, *Chanos chanos, Elops machnata*, *Megalops cyprinoides, Acanthopagras berda,* Hilsa *(Tenulosa) ilisha* and *Rhinomugil corsula* had almost disappeared and were rarely observed in the catches. All these species reappeared in the lake and contributed to the commercial catch during the post-restoration phase. *Acanthopagrus berda* though has reappeared, is yet to form its fishery.

REFERENCES

Baran, E., 2000. Biodiversity of Estuarine Fish Faunas in West Africa. *Naga*, the ICLARM Quarterly, **23(4):** 4-9.

Biradar, R.S., 1998. Fisheries Statistics. *Course Manual* No. 14, Central Institute of Fisheries Education (ICAR), Mumbai, 229p.

Biswas, K.P., 1995. Ecology and fisheries development in wetlands: A study of Chilika lagoon. 192 p.

Bhatta, K.S., Pattnaik, A.K. and Behera, B.P., 2001. Further Contribution to the Fish Fauna of Chilika Lagoon, a coastal wetland of Orissa. *GEOBIOS*, **28(2-3):** 97-100.

Chaudhuri, B.L., 1916a. Description of Two New Fishes from Chilika Lake. *Rec. Indian Mus.*, **12(3):** 105-108.

Chaudhuri, B.L., 1916b. Fauna of the Chilika Lake : Fish Part I. *Mem. Indian Mus.*, **5(4):** 403-440.

Chaudhuri, B.L., 1916c. Fauna of the Chilika Lake: Fish Part II. *Mem. Indian Mus.*, **5(5):** 441-458.

Chaudhuri, B.L., 1917. Fauna of the Chilika Lake: Fish Part III. *Mem. Indian Mus.*, **5(6):** 491-508.

Chaudhuri, B.L., 1923. Fauna of the Chilika Lake: Fish Part IV. *Mem. Indian Mus.*, **5(11):** 711-736.

Devasundaram, M.P., 1954. A report on the fisheries of the Chilika lake from 1948 to 1950. Orissa Government Publ., 1-34.

Fuseya, R. and Watanabe, S., 1996. Genetic Variability in the Mudcrab Genus *Scylla* (Brachyura: Portunidae). *Fisheries Science.*, **62(5):** 705-709.

Fushimi, H. and Watanabe, S., 1999. Problems in Species Identification of the Mudcrab Genus *Scylla* (Brachyura: Portunidae). *UJNR Technical Report No. 28*. Proc. 28th Aquaculture Panel Symposium, US-Japan Cooperative Programme in Natural Resources, Hawaii, November 10-12, 1999: 9-13.

Gupta, R.A, Mandal, S.K. and Paul, S., 1991. Methodology for Collection and Estimation of Inland Fisheries Statistics in India. Central Inland Capture Fisheries Research Institute (ICAR), Barrackpore, West Bengal. *Bull No. 58* (Revised Edition): 64 p.

Hora, S.L., 1923. Fauna of the Chilika Lake : Fish part V. *Mem. Indian Mus.*, **5(11):** 737-770.

Jones, S. and Sujansinghani, K.H., 1954. Fish and fisheries of the Chilika lake with statistics of fish catches for the years 1948-1950. *Indian J. Fish.*, **(1-2):** 256-344.

Jhingran, V.G. and Natarajan, A.V., 1966. Final Report on the Fisheries of the Chilika Lake (1957-1965). Cent. Inl. Fish. Res. Inst., *Bulletin*, **8:** 1-12.

Jhingran, V.G. and Natarajan, A.V., 1969. A study of the fisheries and fish populations of the Chilika lake during the period 1957-65. *J. Inland Fish. Soc. India*. **1:** 49-126.

Kemp, S., 1915. Crustacea: Decapoda Fauna of the Chilika Lake. *Mem. Indian Mus.*, **5:** 199-325.

Koumans, F.P., 1941. Gobioid Fishes of India. *Mem. Indian Mus.*, **13(3):** 205-313.

Khora, S.S., 2002. Icthyofaunal Scenario of Chilika Lagoon. International Workshop on Restoration of Chilika Lagoon (*Abstr. Souv*), Bhubaneswar, 18-20 January, 2002.

Mayadeb, 1995. Crustacea: Brachyura. Wetland Ecosystem Series 1: Fauna of Chilika Lake. *Zool. Surv. India*, 345-366.

Mitra, G.N., 1946. Development of Chilika Lake, Orissa. Govt. Press, Cuttack.

Menon, M.A.S., 1961. On a collection of fish from lake Chilika, Orissa. *Rec. Indian Mus.*, **59(1&2):** 41-69.

Misra, K.S., 1962. An aid to the identification of the common commercial fishes of India and Pakisthan. *Rec. Indian Mus.*, **57(1-4):** 1-320.

Misra, K.S., 1969. Pisces: The Fauna of India and Adjacent Countries. Vol. 1 (2nd ed.) Manager Publication, New Delhi, 276 p.

Misra, K.S., 1976a. Pisces: The Fauna of India and Adjacent Countries. Vol. II (2nd ed.), Manager Publication, New Delhi, 438 p.

Misra, K.S., 1976b. Pisces: The Fauna of India and Adjacent Countries. Vol. III (2nd ed.), Manager Publications, New Delhi, 367 p.

Mohanty, S.K., 1973. Further Additions to the Fish Fauna of the Chilika Lake. *J. Bombay Nat. Hist. Soc.*, **72(3):** 863-866.

Mohanty, S.K., 2002. Fisheries Biodiversity of Chilika Lagoon. *Chilika Newsletter.* January (2002), 11-12.

Mohanty, S.K., 2003. Evaluation of Commercial Fish Landings from Chilika Lagoon before and after Hydrological Intervention. *Chilika Newsletter*, **4:** 20-21.

Mohanty, S.K., Mohanty, Rajeeb K. and Badapanda, H.S., 2001. Fishery Dynamics and Management of the Chilika Lagoon. *Fishing Chime.*, **22(5):** 42-43.

Pattnaik, A.K., 2000. Conservation of Chilika – An overview. *Chilika Newsletter*, **1:** 3-5.

Pattnaik, A.K., 2001. Hydrological Intervention for Restoration of Chilika Lagoon. *Chilika Newsletter*, **2:** 3-5.

Rajan, S., Pattanaik, S. and Bose, N.C., 1968. New records of fishes from Chilika lake. *J. Zool. Soc. India.*, **20(1):** 80-83.

Rama Rao, K.V., 1995. Pisces: Wetland Ecosystem Series 1: Fauna of the Chilika Lake. *Zool. Surv. India,* 483-506.

Reddy, K.N., 1995. Crustacea: Decapoda, Wetland Ecosystem Series 1: Fauna of Chilika Lake. *Zool. Surv. India*, 367-389.

Roy, J.C. and Sahoo, N., 1957. Additions to the Fish Fauna of the Chilika Lake. *J. Bombay Nat. Hist. Soc.*, **54:** 943-953.

Ritesh, K., 2003. Economic valuation of Chilika lagoon. *Chilika Newsletter*, **4:** 24-26.

Talwar, P.K. and Kacker, R.K., 1986. Commercial Sea Fishes of India. Handbook No. 4. *Zool. Souv. India*. Calcutta., 997 p.

Talwar, P.K. and Jhingran, A.G., 1991. Inland Fishing of India and Adjacent Countries. Oxford & IBH, New Delhi, Vol. I & II, 1077 p.

Trisal, C.L., 2000. Sustainable Development and Biodiversity Conservation of Chilika Lagoon. *Chilika Newsletter*, **1:** 6-8.

Biodiversity Assessment of Algae in Chilika Lake, East Coast of India

J. Rath and S.P. Adhikary

Post Graduate Departments of Botany & Biotechnology
Utkal University, Bhubaneswar - 751004

1. INTRODUCTION

Chilika is the largest brackish water lagoon in Asia situated in the east coast of India between 19°28′ and 19°54′ N latitude and 85°06′ and 85°35′ E longitude. The lagoon is an estuarine one and supports an unique assemblage of marine, brackish water and freshwater species. Algal flora of Chilika lake has been studied several times during the last century. Most of these works were repetitive in nature and none of the authors have studied the algae of the lagoon in every season covering the entire catchment area in a particular year. Further, there is also no report available containing information about all the algal forms including both macro- and micro-algal species occurring in the Chilika lake during a particular time covering a year, and also no detail taxonomic account of each of the species available. Therefore, the algal forms occurring throughout the lake in different seasons for two consecutive years were surveyed with a view to study the different algal forms occurring in the lake. Macro-algae as well as phytoplankton were collected in several collection trips during 2000-2001, analyzed and an authentic algal distribution map of the lake was prepared.

2. MATERIALS AND METHODS

Study of algal diversity in Chilika lake was carried out for two consecutive years during 2000 and 2001. All the intertidal regions along the shore line including periphery of the islands, rocks, pebbles, logs and fishing nets etc. of all the four sectors were surveyed in all the three seasons. Macroscopic algal samples were immediately preserved after collection in 4% formalin on the spot. Micro-algae were collected from 24 stations covering all the four sectors of the lake (Fig. 1) at regular intervals using a 10 mm and 20 mm mesh size KC Denmark phytoplankton net. These were not preserved using any preservatives, as they are quite sensitive to these chemicals leading to bleaching of pigments. Thus they were placed in an ice carrier and analyzed soon after reaching the laboratory. Sample number was given to each species following the first letter "C" as Chilika lagoon, the sectors (S, Ce, N, O) as, Southern, Central, Northern and Outer channel sectors respectively, then the first letter of the collection site, sample no. and date of collection in sequence.

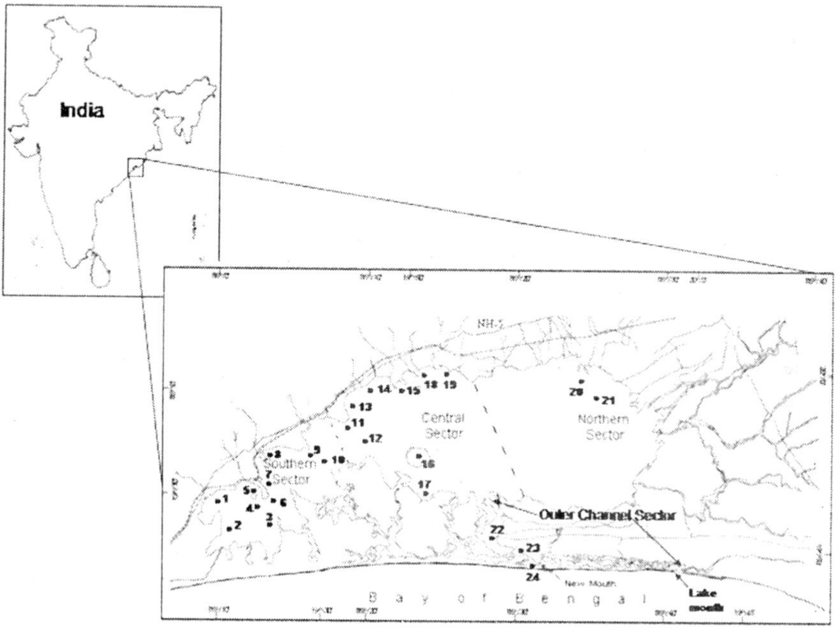

Figure 1: Map of Chilika lake with station locations.

Herbarium of all the macro-algae was prepared and kept in the herbarium of the P.G. Department of Botany, Utkal University, Bhubaneswar. Photographs of all the species and microphotographs of the microscopic forms were taken in a Meiji ML-TH-05 trinocular research microscope fitted with F-50 Nikon camera. Camera lucida diagram of each of the microscopic forms and hand drawing in case of macro-algal form were drawn. Measurement of length and breadth of the microscopic forms was recorded with standard Erma micrometers. The morphological features of each species was compared with the standard keys and literature available in monographs and research publications (Biswas, 1932; Subrahmanyan, 1946; Prescott, 1954; Desikachary, 1959; Paragallo, 1908; Desikachary, 1989; Desikachary et al., 1990, 1998; Cox, 1996; Thomas, 1997 and Krishnamurthy, 1999) and identification was made up to species level. All the species encountered in the study were listed with correct author citation and family name. Most commonly used synonyms and/or names appearing in recent literature have been cited within parenthesis after the correct botanical names, wherever necessary.

3. RESULTS AND DISCUSSION

Based on the resultant data, a checklist of algae of Chilika lake during the year 2000-2001 has been prepared and presented in Table 1. One hundred and two species of algae encountered in the study belong to five divisions e.g. Cyanophyta, Chlorophyta, Bacillariophyta, Dinophyta and Rhodophyta. The numbers of algal species under these divisions were as follows: Cyanophyta 12 species, Chlorophyta 23 species, Bacillariophyta 58 species, Dinophyta five species and four species of Rhodophyta. The percentage occurrence of different algal groups with respect to total algal taxa has been given in Fig. 2. Of these, the dominant species of algae in the lake which occur in higher quantity were *Chaetomorpha linum, Enteromorpha intestinalis, Enteromorpha compressa, Lyngbya aestuarii, Ulva lactuca, Cladophora glomerata, Gracilaria verrucosa, Polysiphonia subtilissima* and *Grateloupia filicina* (Plate I). In addition, many planktonic species occurring in higher numbers and widely were also reported. These were *Coscinodiscus centralis, Odentella mobiliensis, Chaetoceros diversus, Cylindrotheca closterium, Nitzschia obtuse, Amphora ovalis, Thallassionema nitzschioides, Navicula salinarum, Rhizosolenia setigera, Pleurosigma normanii, Gyrosigma acuminatum, Synedra ulna, Pinnularia alpine, Asterionellopsis glacialis, Stauroneis pusilla, Melosira borrerii, Dinophysis caudate* and *Ceratium longipes* (Plate II). The detail taxonomic account with a key to species, photographs and camera lucida diagrams are published elsewhere (Rath and Adhikary, 2005).

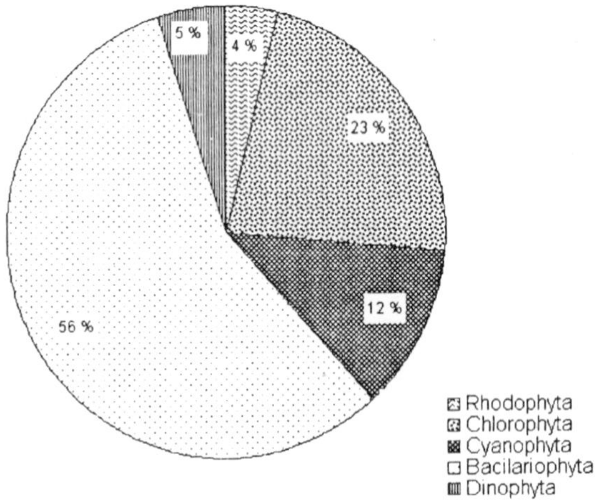

Figure 2: Percentage of different algae occurring in Chilika lake.

Plate I.

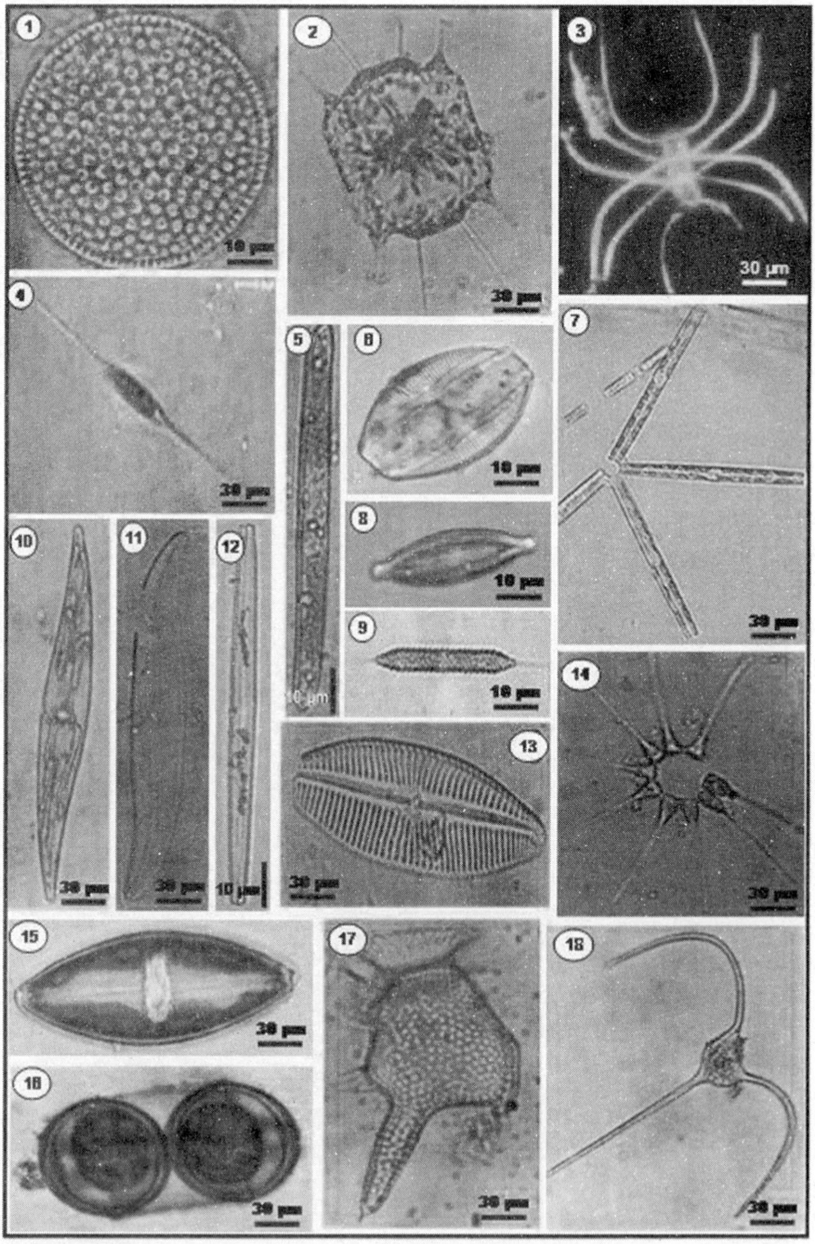

Plate II.

Table 1: Checklist of algae of Chilika lake during 2000-2001

Sl. No.	*Algal species*	*Family*	*New Report in the present study*
Cyanophyta			
1	*Synechocystis aquatilis* Sauv.	Chroococcaceae	
2	*Chroococcus turgidus* Naeg.	Chroococcaceae	+
3	*Merismopedia glauca* (Ehrenb.) Nag.	Chroococcaceae	+
4	*M. elegans* A.BR.	Chroococcaceae	+
5	*Spirulina subtilissima* Kütz. ex. Gomont	Oscillatoriaceae	+
6	*Arthrospira platensis* (Nordst.) Gomont	Oscillatoriaceae	+
7	*Oscillatoria princeps* Vaucher ex. Gomont	Oscillatoriaceae	+
8	*Lyngbya aestuarii* Liebm.	Oscillatoriaceae	
9	*Phormidium submembranaceum* (Ardissone & Strafforells) Gomont	Oscillatoriaceae	
10	*Anabaena torulosa* Lagerh	Nostocaceae	
11	*A. flos-aquae* (Lyngb.) Breb.	Nostocaceae	+
12	*Fischerella* sp.	Stigonemataceae	+
Chlorophyta			
13	*Eudorine elegans* Ehrenberg	Volvocaceae	+
14	*Scenedesmus quadricauda* (Turpin) Brébisson	Scenedesmaceae	+
15	*S. acuminatus* (Lagerheim) Chodat	Scenedesmaceae	+
16	*Coelastrum cambricum* Archer var. *Intermedium* (Bohlin) G.S. West	Coelastraceae	+
17	*Actinastrum hantzschii* Lagerheim	Selenastraceae	+
18	*Selenastrum gracile* Reinsch	Selenastraceae	+
19	*Pediastrum simplex* Meyen var. *duodenarium*(Bailey) Rabenhorst	Hydrodictyaceae	+
20	*P. duplex* Meyen var. *subgranulatum* Racib	Hydrodictyaceae	+
21	*P. duplex* Meyen var. *reticulatum* Lagerheim	Hydrodictyaceae	+
22	*P. tetras* (Her.) Ralfs.	Hydrodictyaceae	+
23	*Tetraedron gracile* (Reinsch) Hansgirg	Hydrodictyaceae	+
24	*T. trigonum* (Naegeli) Hansgirg	Hydrodictyaceae	+
25	*Xanthidium sexmamillatum* W & G.S.West	Zygnemaceae	+
26	*Cosmarium impressulum* Elfv.	Zygnemaceae	+
27	*Spirogyra* sp.	Zygnemaceae	

(*contd.*)

Table 1 (*contd.*)

Sl. No.	*Algal species*	*Family*	*New Report in the present study*
28	*Cladophora glomerata* (Linn.) Kutzing	Cladophoraceae	
29	*Chaetomorpha linum* (O.F.Muller) Kutzing	Cladophoraceae	
30	*Enteromorpha intestinalis* (Linn.) Neesa	Ulvaceae	
31	*E. compressa* (Linn.) Nees	Ulvaceae	
32	*E. usneoides* (Bonnem.) J. Ag.	Ulvaceae	+
33	*Ulva lactuca* Linn.	Ulvaceae	
34	*Chara* sp.	Characeae	
35	*Nitella* sp.	Characeae	
Bacillariophyta			
36	*Coscinodiscus marginatus* Ehrenberg	Coscinodiscaceae	+
37	*C. gigas* Ehrenberg	Coscinodiscaceae	+
38	*C. centralis* Ehrenberg	Coscinodiscaceae	
39	*Coscinodiscus* sp. I	Coscinodiscaceae	
40	*Coscinodiscus* sp. II	Coscinodiscaceae	
41	*Auliscus sculptus* (W. Smith) Ralfs.	Eupodiasceae	+
42	*Thalassiosira subtilis* (Ostenfeld) Gran.	Thalassiosiraceae	
43	*Skeletonema costatum* (Greville) Cleve	Thalassiosiraceae	
44	*Lauderia annulata* Cleve	Thalassiosiraceae	
45	*Stephanopyxis turris* (Grev.et Arn.) Ralfs.	Melosiraceae	
46	*Melosira borreii* Grev.	Melosiraceae	+
47	*Guinardia flaccida* (Castracane) H. Peragallo	Rhizosoleniaceae	+
48	*Rhizosolenia setigera* Brightwell	Rhizosoleniaceae	
49	*Leptocylindrus danicus* Cleve	Leptocylindraceae	+
50	*Odentella mobiliensis* (Bailey) Grunow (Synonym- *Biddulphia mobiliensis* Grunow)	Eupodiscaceae	
51	*Biddulphia heteroceros* Grunow	Biddulphiaceae	
52	*Biddulphia* sp.	Biddulphiaceae	
53	*Ditylum brightwellii* (West) Grunow	Lithodesmiaceae	
54	*Bacteriastrum hyalinum* Lauder	Chaetoceraceae	
55	*B. furcatum* Cleve	Chaetoceraceae	+
56	*Chaetoceros curvisetus* Cleve	Chaetoceraceae	
57	*C. eibenii* Grunow	Chaetoceraceae	
58	*C. diversus* Laud. var. *tenuis* Cleve	Chaetoceraceae	

(*contd.*)

Table 1 (*contd.*)

59	*C. affinis* Lauder	Chaetoceraceae	
60	*C. lorenzianus* Grunow	Chaetoceraceae	
61	*C. paradoxus* Cleve	Chaetoceraceae	+
62	*Chaetoceros* sp.	Chaetoceraceae	
63	*Synedra ulna* (Nitzsch) Ehr. var. *Danica* (Kütz.) Grun.	Fragilariaceae	+
64	*Diatoma elongatum* Agardh	Fragilariaceae	+
65	*Lichmophora abbreviata* Agardh	Fragilariaceae	+
66	*Climacosphaenia moniligera* Ehrenberg	Fragilariaceae	+
67	*Asterionellopsis glacialis* (Castracane) Round (Synonym- *Asterionella japonica* Cleve)	Fragilariaceae	
68	*Grammatophora undulata* Ehrenberg	Fragilariaceae	+
69	*Fragilaria crotonensis* Kitton	Fragilariaceae	+
70	*Fragilaria* sp.	Fragilariaceae	+
71	*Thallassionema nitzschioides* Grunow	Thallassionemataceae	
72	*Stauroneis pusilla* A.Cleve	Naviculaceae	+
73	*Pinnularia alpina* W. Smith	Naviculaceae	+
74	*P. nobilis* Ehrenberg	Naviculaceae	+
75	*Amphora ovalis* Kützing	Naviculaceae	+
76	*Amphiprora gigantean* Grunow	Naviculaceae	+
77	*Amphora* sp.	Naviculaceae	+
78	*Navicula protracta* (Grunow) Cleve.	Naviculaceae	+
79	*N. minuscule* Grunow	Naviculaceae	+
80	*N. lanceolata* (C.A. Agardh) Kützing	Naviculaceae	+
81	*N. salinarum* Grunow	Naviculaceae	+
82	*Navicula* sp.	Naviculaceae	
83	*Pleurosigma normanii* Ralfs.	Naviculaceae	+
84	*Gyrosigma acuminatum* (Kutz.) Rab.	Naviculaceae	+
85	*Cymbela* sp.	Naviculaceae	+
86	*Craticula cuspidate* (Kützing) D.G. Mann	Naviculaceae	+
87	*Nitzschia obtusa* W. Smith.	Bacillariaceae	+
88	*N. pandoriformis* Gregory	Bacillariaceae	+
89	*N. sigma* (Kützing) W. Smith	Bacillariaceae	+
90	*Cylindrotheca closterium* (Ehrenberg) Lawin & Reimann. (Synonym: *Nitzschia closterium*)	Bacillariaceae	
91	*Bacillaria paxillifera*		

(*contd.*)

Table 1 (*contd.*)

Sl. No.	*Algal species*	*Family*	*New Report in the present study*
	(O.F. Muller) Hendey (*Bacillaria paradoxa* Gmelin)	Bacillariaceae	
92	*Cocconeis pediculus* Ehrenberg	Achnanthoideae	+
93	*Tabellaria fenestrata* (Lyngbye) Kützing	Tabellariaceae	
Dinophyta			
94	*Ceratium lineatum* (Ehrenberg) Cleve	Ceratiaceae	+
95	*C. tripos* (O.F.Müller) Nitzsch	Ceratiaceae	
96	*C. longipes* (Bailey) Gran.	Ceratiaceae	
97	*Dinophysis caudate* Saville-Kent	Dinophysiaceae	
98	*Gymnodinium heterostriatum* Kofoid & Swezy	Gymnodiniaceae	+
Rhodophyta			
99	*Ceramium diaphanum* (Lightfoot) Roth var. elegans (Roth) Roth. (*Ceramium elegans* Ducl.)	Ceramiaceae	
100	*Polysiphonia subtilissima* Mont	Rhodomelaceae	
101	*Gracilaria verrucosa* (Hudson) Papenfuss	Gracilariaceae	
102	*Grateloupia filicina* (Wulf.) Ag	Hylymeniaceae	

Algal flora of Chilika lake was first studied as early as in 1930's by Biswas. During these last seventy years only a few published work on algal flora of Chilika lake is available. Of these, only four reports were on both the macro- and micro-algal forms occurring in all sectors of the lake. These were of Biswas (1932), Ahmed (1966), Patnaik (1978) and Adhikary and Sahu (1992). Three publications during 1954 to 1990, e.g. Roy (1954), Patnaik (1973) and Raman et al. (1990) only dealt with the phytoplankton of Chilika lake. Sahu and Adhikary (1999) covering ten years of study during 1990-1999 gave a picture of certain widely occurring macro-algal forms in Chilika lake. Analysis of these data together with the algal forms encountered in the Chilika lake during 2000-2001 showed that those macro-algal forms reported by Biswas in 1932 occurred in the lake for sometime up to 1969 and after that many species such as *Polysiphonia sertularioides* and *Gracilaria lichenoides* disappeared from the lake. Many of the commonly growing species such as *Ceramium diaphanum, Gracilaria verrucosa, Grateloupia filicina, Phormidium submembranaceum, Chaetomorpha linum, Cladophora glomerata, Entromorpha compressa, Enteromorpha intestinalis* and *Lyngbya aesturii* were found occurring in the lake since last seventy years and, hence, are the major forms contributing to the algal biomass of Chilika lake. *Chara* sp., *Nitella* sp. and *Ulva lactuca,* which were not reported by Biswas in 1932, were detected in the lake afterwards. Similarly, few blue green algae such as *Lyngbya aerugineo-caerulea, Lyngbya confervoides* and *Microcoleus* sp. and certain *Oscillatoria, Phormidium* and *Rhizosolenia* species reported by Biswas in 1932 disappeared from the lake during the subsequent years. As regards the distribution of Bacillariophytes and phytoplankton species in the lake reported by Roy (1954), Ahmed (1969), Patnaik (1973) and Raman (1990), no distinct information emerged from the comparative account of the reports over the years between 1954 and 1990. These published data were quite contradictory to each other; hence no clear conclusive information could be gathered analyzing these publications. A detailed study undertaken by the authors (Rath and Adhikary, 2005) on the algal forms of Chilika lake documented description to each species, their micro-photograph, camera lucida diagrams and systematic enumeration of each form. The total number of algal forms (both micro- and macro-algal forms) reported by Biswas (1932a), Ahmed (1966), Patnaik (1978) and Adhikary and Sahu (1992) were 22, 12, 21 and 28 respectively, whereas in the present study these forms recorded were 102 in number. As regards phytoplankton, their total number in Chilika lake reported by Roy (1954), Patnaik (1973) and Raman et al. (1990) were 33, 57 and 19 respectively. However, in the present study the number of phytoplankton species (except macro-algae) were 84 in number. Raman et al. (1990) though mentioned that they have found 97 species of phytoplankton in the lake belonging to Cyanophyta, Chlorophyta, Bacillariophyta and Dinophyta, but had mentioned the name of only 19 species. Since neither description of these species nor their identity was mentioned in the publication, the exact Phytoplankton species they have encountered and mentioned in the paper was not considered for the comparison.

4. CONCLUSIONS

In the present study total 58 species of algae not previously reported belonging to four divisions were recorded from the lake. Of these, eight species belong to Cyanophyta, 15 species to Chlorophyta, 33 species to Bacillariophyta and two species to Dinophyta group (Table 1). Of the three species of *Enteromorpha, E. intestinalis* as well as *E. compressa* have been reported in the Chilika lake by several authors since the first report of Biswas (1932). However, in the present study another species of *Enteromorpha* viz. *E. usneoides* was observed in the Southern sector of the lake for the first time, which is new distribution range for this species in India. This species is characterized by its well developed branching patterns (Plate III), and is quite different from *Enteromorpha intestinalis* and *Enteromorpha compressa.* Bliding (1963) described this species as *Enteromorpha compressa* var

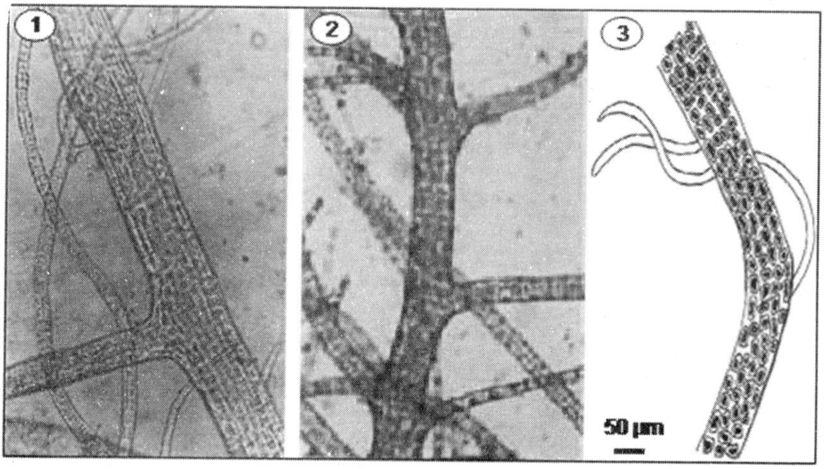

Plate III.

usneoides but Koeman and Hoek (1982) described the organism from Netherlands as *Enteromorpha usneoides*.

ACKNOWLEDGEMENTS

We thank Dr. A.K. Pattnaik, Chief executive, Chilika Development Authority for providing facility for sample collection and financial assistance to one of us (JR). Thanks are also due to the Head of the Department of Botany, Utkal University, Bhubaneswar for laboratory facilities.

REFERENCES

Ahmed, M.K., 1966. Studies on *Gracilaria* Greu of the Chilika lake. *Bull. Orissa Fish Res. Invest.* **1:** 46-53.

Adhikary, S.P. and Sahu, J.K., 1992. Distribution and seasonal abundance of Algal forms in Chilika lake, East coast of India. *Jpn. Limnol.* **53:** 197-205.

Biswas, K., 1932. Algal flora of the Chilika lake. *Mem Asiat Soc. Bengal.* **11:** 65-198.

Bliding, C.A., 1963. Critical survey of European taxa in Ulvales. Part 1. *Capsosiphon, Percursaria, Blidingia, Enteromorpha. Opera Botanica Lund.* **8:** 1-160.

Cox, J.E., 1996. *Identification of Freshwater Diatoms from Live Material.* Chapman & Hill London. pp. 158.

Desikachary, T.V., 1959. *Cyanophyta.* Indian Council of Agricultural Research, New Delhi, India. pp. 686.

Desikachary, T.V., 1989. *Atlas of Diatoms.* Monographs. Fasicle II, III and IV. Madras Science Foundation, Madras.

Desikachary, T.V., Krishnamurthy, V. and Balakrishan, M.S., 1990. *Rhodophyta.*

Madras Science Foundation Madras. Vol. I, pp. 548.

Desikachary, T.V., Krishnamurthy, V. and Balakrishan, M.S., 1998. *Rhodophyta*. Madras Science Foundation Madras. Vol. II Part II B, pp. 359.

Koeman, R.P.T. and Hoek C.V.D., 1982. The taxonomy of *Enteromorpha* Link, (Chlorophyceae) in the Netherlands. *Arch. Hydrobiology. Supp. 63, Algological Studies*. **32:** 279-330.

Krishnamurthy, V., 1999. *Algae of India and neighbouring countries I. Chlorophycota*. Oxford & IBH, New Delhi. pp. 209.

Patnaik, S., 1973. Observation on the seasonal fluctuating of plankton in the Chilika lake. *Ind. J. Fish*, **20:** 43-45.

Patnaik, S., 1978. Distribution and Seasonal abundance of some algal forms in Chilika lake *J. Inl. Fish Soc. India*, **10:** 56-67.

Paragallo, H. and Peragallo, M., 1908. Diatomees Marines de France (M.J. Tempere ed.) Grez-Sur-Loing. pp. 492, I-XII, 138 plates.

Prescott, G.W., 1954. How to know the fresh water Algae. WM.C Brown Company, Dubuque, Iowa, pp. 211.

Rath, J. and Adhikary, S.P., 2005. Algal flora of Chilika lake. Daya Publishing House, New Delhi.

Raman, A., Satyanarayana, Ch., Adiseshasai, K. and Prakash, K.P., 1990. Phytoplankton characteristic of Chilika lake, a brackishwater lagoon along the east coast of India. *Indian J. Mar. Sci.*, **19:** 274-277.

Roy, J.C., 1954. Periodicity of plankton diatoms of the Chilika lake for the years 1950-51. *J. Bom. Natl. Hist. Soc.*, **52:** 112-123.

Sahu, J.K. and Adhikary, S.P., 1999. Distribution of seaweeds in Chilika lake. *Seaweed Res. Utiln.* **21:** 55-59.

Subrahmanyan, R., 1946. A systematic account of the marine planktonic diatoms of Madras coast. *Proc. Indian Acad. Sci.* **24B:** 85-197.

Thomas, C.R., 1997. *Identifying marine phytoplankton*. Academic Press, California, USA. pp. 846.

Mercury Resistant *Bacillus Cereus* Isolated from the Pulicat Lake Sediment, North Chennai Coastal Region, South East India

S. Kamala Kannan, S. Mahadevan[1] and R. Krishnamoorthy

Department of Applied Geology
University of Madras, Chennai - 600 025, India
skk2k@rediffmail.com
[1]Department of Molecular Reproduction, Development and Genetics
Indian Institute of Science, Bangalore - 560 012, India
mahi@mrdg.iisc.ernet.in

1. INTRODUCTION

Among the 90 naturally occurring elements, 21 are non-metals, 16 are light metals and the remaining 53 (with As including) are heavy metals. In these 53 heavy metals not all elements are toxic: some are required for catalyzing key reactions or for maintaining the protein structure, to maintain the osmotic balance and to catalyze several biochemical reactions. Hence, based on the physiological point of view, the metals are classified into three groups namely (i) Essential and basically non-toxic e.g., Ca and Mg, (ii) Essential but harmful under higher concentration e.g., Fe, Mn, Zn, Cu, Co etc. and (iii) Toxic metals e.g., Hg and Cd (Valls and Lorenzo, 2002).

Contamination of the environment with toxic heavy metals is of growing concern because these compounds move up food chain, may get biomagnified and cause numerous acute and chronic health disorders in human beings. Unlike organic contaminants these heavy metals are not readily biodegradable, which can be transformed from one chemical state to another. Therefore, heavy metal pollution in the environment is a raising concern (Barkay and Schaefer, 2001). Hence, the present study was focused on the bio-reduction of toxic heavy metals especially to mercury, because of its high toxicity (0.01 μg/ml), wide application in industries and their ability to persist several decades in aquatic environment (Wagner-Dobler et al., 2000).

Removal of mercury from the contaminated site especially in coastal areas is a challenge for environmental management. Commonly used mercury removal process such as those based on ion exchange or biosorbents have been shown to be sensitive to environmental conditions. Therefore these methods will not be successful to remove and recover the metal ions. Microbial detoxification will be helpful for the development of possible biotransformation process. Basic principle behind the bio-transformation of inorganic mercury (Hg^{2+}) to volatile mercury (Hg^0) is a NADPH mediated, flavin adenine dinucleotide containing disulfide oxide reductase, the mercury reductase, genes encoding the enzyme and other functional proteins for regulation and transport were clustered in an operon called *mer* operon. However, in some cases an additional gene merB found in association with other genes and encodes for organomercurial lyase, which cleaves the carbon mercury bond of organo mercuric compounds and allows the inorganic mercuric compound for subsequent biotransformation into volatile form (Barkay et al., 1990).

2. STUDY AREA

Pulicat lake is the second largest brackish water lagoon in the country, which runs parallel to the Bay of Bengal, bordering the east coast in Nellore district of Andhra Pradesh, with a portion of it extending into Thiruvallur district of Tamil Nadu. The lake is located 40 km north of Chennai (formerly Madras) and is separated from the Bay of Bengal by Sriharikota island. The lake is about 360 sq km in area and its depth (water column) varies from 1 to 6 m. Four rivers—the Swarnamukhi, the Kalangi, the Araniar and the Royyala Kalava—are the major sources for the fresh water flows. Pulicat lake contains diverse natural resources, which include aquatic and terrestrial fauna and flora (Krishnakumar, 2000). The improperly treated industrial effluents, including those from two coal-based power plants and petrochemical industries from the Ennore Creek and Buckingham Canal, ultimately reach the Pulicat lake through its mouth from the Bay of Bengal coastal waters. Earlier studies in Pulicat lake have recorded elevated levels of heavy metal concentrations, especially cadmium, mercury and arsenic (Padma and Periakali, 1998a, 1998b, 1999).

The present study is, therefore, aimed at isolation and identification of mercury reducing bacteria from the heavy metal contaminated Pulicat lake sediments, in order to establish the possibility of employing the isolates for bioremediation of mercury containing industrial waste waters as well as for in situ bioremediation.

3. MATERIALS AND METHODS

3.1 Sampling and Characterization of the Isolates

The surface sediment samples from the Pulicat lake were collected using Petersen grab and transported on ice to the laboratory and enumerated for mercury resistant bacteria within 5-6 hours of sample collection. Samples were serial diluted in sterile distilled water and 0.1 ml of the appropriate dilution was plated by the spread plate technique on Zobel Marine Agar plates (Hi-media India) amended with mercury as mercuric chloride (Srinath, 2002). Later, the plates were incubated at 26° to 28° C for 24 hours and observed for bacterial growth. The isolates were identified based on morphology and biochemical characters.

3.2 Minimal Inhibitory Concentration (MIC) of Mercuric Compounds

Minimal inhibitory concentrations of mercuric compounds were determined by agar dilution method (Luli et al., 1983). The mid-log phase cultures of the isolates were aseptically transferred to the nutrient agar plates (Hi-Media, India) supplemented with different concentrations of mercuric compounds as mercuric chloride and phenyl mercuric acetate (PMA). Later, the plates were incubated at 26° to 28° C for 24 hours and observed for bacterial growth. The lowest concentration of mercuric compounds at which no growth occurred was considered as MIC.

3.3 Minimal Inhibitory Concentration of other Heavy Metals

Mercury resistance isolates that were also tested for their tolerance to other heavy metals like zinc, nickel, lead, copper, chromium, cobalt, and cadmium were determined by agar dilution method (Luli et al., 1983). The fresh overnight cultures of the isolates were aseptically inoculated into the nutrient agar plates, which were supplemented with different concentrations of the aforesaid heavy metals individually (Verma et al., 2001). The metal salts used for the study include K_2CrO_4 (Qualigens, India), $COCl_2.6H_2O$ (CDH, India), $Pb(NO_3)_2$ (CDH, India), CdI_2 (Ranbaxy, India), $NiSO_4.6H_2O$ (Qualigens, India), $ZnCl_2$ (Qualigens, India), $CuSO_4.5H_2O$ (CDH, India), $HgCl_2$ (Reachem, India) and PMA (S.D.Fine, India). The presence of growth after overnight incubation at 26° to 28° C was considered as resistance to the tested metals.

3.4 Growth Kinetics of *Bacillus Cereus* in Presence of Mercuric Compounds and Chromate

To study the effect of mercuric compounds and chromate on the growth of *Bacillus cereus*, the organism was allowed to grow in peptone water (adjusted to pH 7) containing 100 μg/ml and 200 μg/ml of mercuric chloride, 5 μg/ml and 10 μg/ml of phenyl mercuric acetate and 100 μg/ml and 200 μg/ml chromate. The cultures grown in the absence of metals served as a control in this experiment.

Five ml of the mid-log phase cultures of *Bacillus cereus* grown in peptone water were used for inoculation into the Erlenmeyer flasks containing 50 ml of peptone water (Sangeeta and Tripathi, 2001). The flasks were incubated at 28° C and the growth was measured at different time intervals in terms of increase in optical density at 482 nm using a spectrophotometer (Systronics 1304).

3.5 Mercury Volatilization Assay

Mercuric resistant isolate *Bacillus cereus* was assayed for their ability to volatilize mercuric chloride has been performed based on Nakamura and Nakahara (1998) method. Mid log-phase culture of the isolate were harvested and transferred into the micro titer plates containing 50 µl of sodium phosphate buffer, 0.05 mM EDTA, 0.2 mM magnesium acetate with 50 µg/ml mercuric chloride. The plate was incubated for three hours at 30°C covered with X-ray film.

3.6 Chromate Reduction Assay

Reductions of chromate were determined by measuring the optical density at 382 nm using a spectrophotometer (Bopp and Ehrlich, 1988). Optical densities of the chromate in the cell-free filtrates were used as a control. Five ml of the mid-log phase culture of the isolate *Bacillus cereus* was aseptically transferred into the flasks containing 50 ml of peptone water supplemented with 50 µg/ml of chromate as K_2CrO_4. The flasks were incubated at 28° C and the aliquots were taken at different time intervals, and cell-free filtrates (0.45µ pore size) were prepared. Optical density of the cell-free filtrates was measured at 382 nm using a spectrophotometer.

4. RESULTS

4.1 Identification of the Isolates

Preliminary identifications of the isolates were performed based on the Gram reactions and presence of spores. Gram-positive with spores and Gram-negative bacilli were observed and, later, the isolates were identified based on physiological and biochemical characterization such as starch hydrolysis, casein hydrolysis, gelatin liquefaction, nitrate reduction, IMViC, catalase, oxidase and sugar fermentations.

4.2 Minimal Inhibitory Concentrations of Mercuric Compounds and Other Metals

Minimal inhibitory concentrations of mercuric compounds and other heavy metals for the isolates are presented in Table 1. *Bacillus cereus* shows high tolerance to all the heavy metals tested, except copper, to which *Bacillus sphaericus* shows high tolerance. Chromium and lead are less toxic to the isolates for *Bacillus* species when compared with other heavy metals. The order of toxicity of the metals to the isolate *Bacillus cereus* was cadmium > copper > zinc > cobalt > nickel > lead > chromium. In general, when compared with *Escherichia coli* and *Vibrio* species, *Bacillus* species showed more resistance to the heavy metals tested.

Table 1: Minimal inhibitory concentrations of the isolates to various heavy metals in μg/ml

Name of the metal	*Baciilus sphaericus*	*Bacillus pumilus*	*Bacillus cereus*	*Escherichia coli*	*Vibrio mimicus*	*Vibrio anguillarum*
Chromium	1600	1600	1620	300	150	330
Cadmium	20	50	150	0.5	50	10
Nickel	750	550	850	50	120	230
Copper	440	100	350	110	40	110
Mercuric chloride	50	400	550	25	35	20
PMA	5	35	40	5	5	10
Lead	850	1010	1050	110	80	60
Zinc	30	100	420	125	60	25
Cobalt	250	100	350	230	160	255

4.3 Effect of Metals on Cell Growth, Mercury Volatilization and Chromate Reduction

The growth pattern of *Bacillus cereus* in the presence of increasing concentrations of chromate, mercuric chloride and phenyl mercuric acetate (Figs 1, 2 and 3) did not follow the same pattern as the control, indicating that the growth rate of the isolate was altered according to increasing concentrations of heavy metals. It is difficult to make a comparison of our results with other literature because the growth rate depends upon several factors like size of the inoculum, temperature, composition of the medium and availability of the metals.

Mercury resistant isolate *Bacillus cereus*, also able to transform the soluble form (Hg^{2+}) of mercury into volatile form (Hg^0), has been observed based on the formation of clear zone in the X-ray film. The mercury resistant *Bacillus cereus* also exhibited the ability to reduce chromate upto 52.3% (Fig. 4) in 48 hours with reference to the original concentration in the medium.

5. DISCUSSION

Three mercury resistant *Bacillus* species, two *Vibrio* species and *Escherichia coli* were isolated from the heavy metal contaminated Pulicat lake sediments. Among the isolates, *Bacillus* species showed more resistance to almost all the heavy metals tested, which is similar to the results of other authors (Ross et al., 1981; Konopka et al., 1999; Viti et al., 2001), who widely documented the presence of *Bacillus* species in the heavy metal contaminated soils.

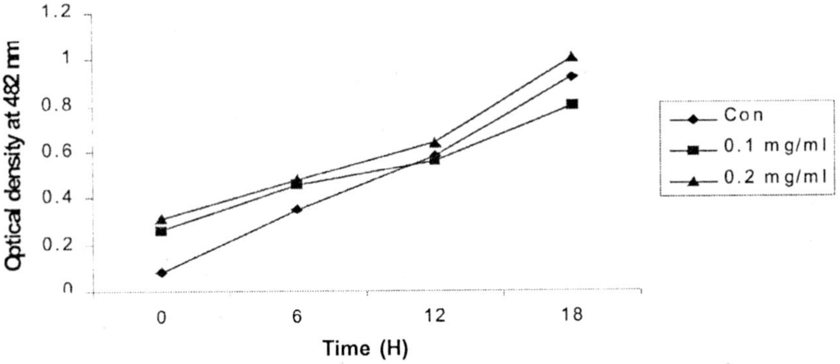

Figure 1: Growth of *Bacillus cereus* in presence of chromate.

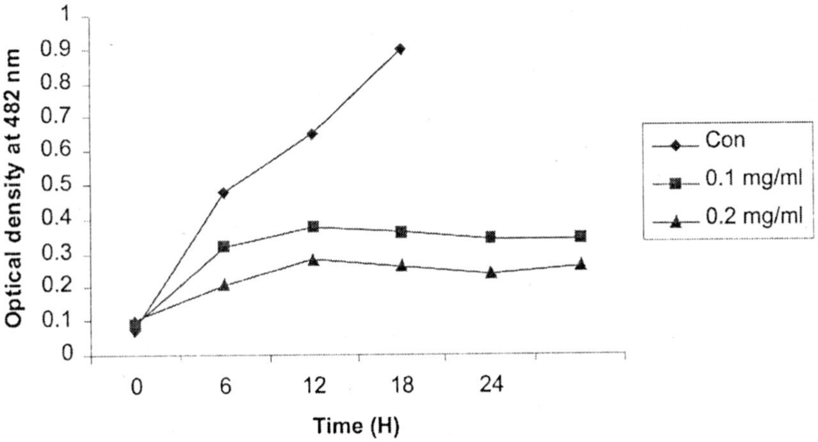

Figure 2: Growth of *Bacillus cereus* in presence of mercuric chloride.

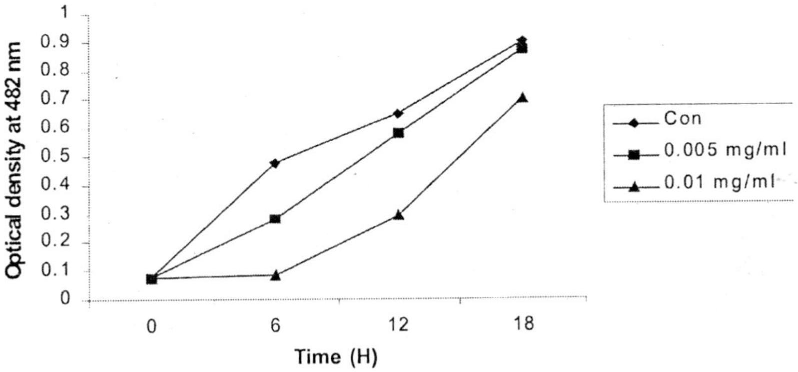

Figure 3: Growth of *Bacillus cereus* in presence of phenyl mercuric acetate.

Among the *Bacillus* species, *Bacillus cereus* showed high levels of tolerance to mercuric chloride and phenyl mercuric acetate (550 μg/ml and 40 μg/ml) in plate test. Since the minimal inhibitory concentration to the above-mentioned heavy metals was determined in solid media, the direct comparison of our data with others could not be attempted because of various parameters like (i) type of the medium used, (ii) composition of the medium, and (iii) availability of metals to the bacteria (Viti et al., 2003) may influence the level of minimal inhibitory concentration.

In the case of growth kinetics, it was observed that the presence of increasing concentrations of chromate, mercuric chloride and phenyl mercuric acetate altered the growth rate of the isolate with increasing concentrations (Figs 1, 2 and 3), which confirms the general decrease in the growth in the increasing concentration of heavy metals. *Bacillus cereus* grown in a medium containing 10 μg/ml of phenyl mercuric acetate showed a much longer lag phase than the control. Results of the chromate reduction indicate that the isolate *Bacillus cereus* has the ability to reduce chromate i.e., it converts the highly toxic chromium (VI) into less toxic chromium (III). It has been noted that all the chromium resistant organisms do not have the ability to reduce chromate (Viti et al., 2003). In our case, all the strains belonging to *Bacillus* species had the ability to reduce chromate (data not shown for *B. pumilus* and *B. sphaericus*).

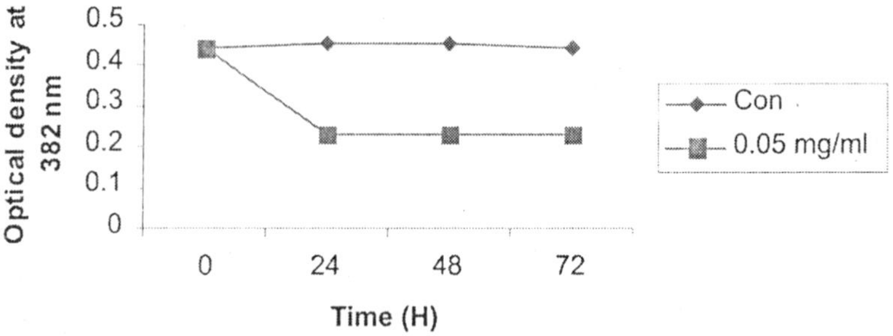

Figure 4: Chromate reduction of *Bacillus cereus*.

6. CONCLUSIONS

Since the industrial effluents are continuously released without proper treatment into the North Chennai coastal region these would not only affect the growth rate and survival of the bacteria, but also affect the biogeochemical cycle of mercury and finally the ecosystem. Based on the results, it is suggested that the presence of mercury-reducing *Bacillus cereus* in the sediments of Pulicat lake has raised the possibility of using this microorganism for bioremediation of mercury-polluted zones. The advantage of selecting indigenous bacteria from the contaminated environment may minimize the inhibitory effects caused by other major chemicals and the physicochemical conditions present in that environment. In addition to efficient mercury and chromate reduction and resistance to multiple heavy metals under aerobic conditions, the other advantages of this indigenous bacterial bioremediation lie in the fact that it is inexpensive, does not require high energy, does not release any additional chemicals into the environment and, above all, is non-hazardous. Hence this study recommends usage of the isolate *Bacillus cereus* for the treatment of metal-based industrial wastewaters containing several metal ions. This study is expected to provide more benefits towards the recovery of degraded coastal environments for the benefit of local community by enriching the resources potential.

ACKNOWLEDGEMENTS

The author (SKK) is thankful to Prof. P. Periakali, Head, Department of Applied Geology, University of Madras, Chennai for his constant encouragement in carrying out this study.

REFERENCES

Barkay, T., Gillman, M. and Liebert, C., 1990. Genes encoding mercuric reductase from selected Gram – negative aquatic bacteria have a low degree of homology with *merA* of Transposon Tn*501*. *Applied and Enviornmental Microbiology*, **56(6):** 1695-1701.

Barkay, T. and Schaefer, J., 2001. Metal and radionuclide bioremediation: Issues, considerations and potentials. *Current Opinion in Microbiology*, **4:** 318-323.

Bopp, L.H. and Ehrlich, H.L., 1988. Chromate resistance and reduction in *Pseudomonas fluorescens* strain LB 300. *Archie Microbiology*, **150:** 426-431.

Konopka, A. and Zakarova, T., 1999. Quantification of bacterial lead resistance via activity assays. *Journal of Microbiological Methods*, **37:** 17-22.

Krishnakumar, A., 2000. Pulicat in peril. *Frontline* 17, Issue 12.

Luli, G.W., Talnagi, J.W., Strohl, W.R., Pfister, R.M., 1983. Hexavalent chromium resistant bacteria isolated from river sediments. *Applied Environmental Microbiology*, **46:** 846-854.

Nakamura, K. and Nakahara, H., 1988. Simplified X-ray film method for the detection of mercurial volatilization of mercuric chloride by *Escherichia coli*. *Applied and Environmental Microbiology*, **545:** 2871-2873.

Padma, S. and Periakali, P., 1998. Cadmium in Pulicat lake sediments, east coast of India. *Environmental Geochemistry,* **1(2):** 55-58.

Padma, S. and Periakali, P., 1998. Mercury in Pulicat lake sediments, East Coast of India. *Journal of Indian Association of Sedimentologists*, **17(2):** 239-244.

Padma, S. and Periakali, P., 1999. Seasonal variation of arsenic in Pulicat lake sediments, east coast of India. *Indian Journal of Environmental Protection*, **19(2):** 125-131.

Ross, D.S., Sjogren, R.E. and Barlett, R.J., 1981. Behavior of chromium in soils: IV. Toxicity to microorganisms. *Journal of Environmental Quality*, **2:** 145-168.

Sangeeta, K. and Tripathi, A.K., 2001. Reduced accumulation of Chromium confers chromate tolerance in *Pseudomonas aeruginosa* from tannery effluent. *Indian Journal of Microbiology*, **41:** 311-313.

Srinath, T., Verma, T., Ramteke, P.W. and Garg, S.K., 2002. Chromium (VI) biosorption and bioaccumulation by chromate resistant bacteria. *Chemosphere*, **48:** 427-435.

Verma, T., Srinath, T., Gadpayle, R.U., Ramteke, P.W., Hans, R.K. and Garg, S.K., 2001. Chromate tolerant bacteria isolated from tannery effluent. *Bioresources technology*, **78:** 31-35.

Valls, M. and Victor De Lorenzo., 2002. Exploiting the genetic and biochemical capacities of bacteria for the remediation of heavy metal pollution. *FEMS Microbiology Reviews*, **26:** 327-338.

Viti, C. and Giovannetti, L., 2001. The impact of chromium contamination on soil heterotrophic and photosynthetic microorganisms. *Annual Microbiology*, **51:** 201-213.

Viti, C., Pace, A. and Giovannetti, L., 2003. Characterization of Cr (VI)- resistant bacteria isolated from the chromium-contaminated soil by Tannery activity. *Current Microbiology*, **46:** 1-5.

Wagner-Dobler, I., Harald Von Canstein, Li, Y., Timmis, K.N. and Deckwer, W., 2000. Removal of mercury from chemical wastewater by Microorganisms in technical scale. *Environmental Science Technology*, **34(21):** 4628-4634.

Mercury Pollution in Vembanadu Lake and Adjoining Muvattupuzha River, Kerala, India

Mahesh Mohan and P.K. Omana

Chemical Science Division, Centre for Earth Science Studies
Akkulam, Thiruvananthapuram
mahim80@rediffmail.com

1. INTRODUCTION

Heavy metals are highly poisonous when they get accumulated or undergo biomagnification processes. The long existence of heavy metals in the aquatic environment is obtained by the bonding with sediments. Among those heavy metals, mercury is considered as the most dangerous pollutant to natural environment because of the ability of plants and animals to accumulate it (Porvari and Verta, 2003) and because of its detrimental effects even at very low concentration (Nriagu, 1979). Mercury present in air and water has increased dramatically in the last century owing to anthropogenic activities. Recent studies suggest that the total global atmospheric mercury has increased between 200 and 500 percent since the beginning of the Industrial Age (UNEP, 2002). Reports also indicate that its levels in rivers, coastal waters, and soil and food items are well above the acceptable levels especially in developing countries like India (Toxics Link, 2003). Pulp and paper, chlor-alkali, and other industries have used mercury for various purposes and it was prescribed that they were losing some of it in their wastewater. The losses of mercury from paper and pulp industry are without doubt a serious problem because much mercury is incorporated in the fibers that settle and remains in the sediments downstream (Hanson, 1971). Fertilizers, pesticides and fungicides are also a source of mercury pollution.

The coastal zone represents one of the earth's complex and dynamic ecosystems such as estuaries and backwaters. In these areas which fringe the continents of the globe, the intricate terrestrial and marine systems become ever more complex as they respond to the pressure created by man (UNESCO, 1976). The estuaries of Kerala, which lie in the southern corner of peninsular

India are exception in the sense that a number of rivers open into a single estuary through backwaters or lakes. Vembanadu lake is the largest brackish, humid tropical wetland ecosystem on the southwest coast of India. This lake system is fed by 10 rivers and is one of the Ramsar sites of India. The Vembanadu lake is connected with Arabian sea through the Cochin estuary. Not only the tidal effect but also the fresh water input during the southwest monsoon is affecting the water quality of the lake. Muvattupuzha river is discharging a large amount of the agricultural waste and effluents from paper and pulp industry into the central part of the lake. No systematic study was conducted to assess mercury pollution in this area.

The present study deals with the contamination in the bed sediment and surface water on a part of the Vembanadu lake with respect to point sources of pollution at Muvattupuzha river.

2. MATERIALS AND METHODS

Vembanadu estuary situated between 76°15′E and 76°25′E to 9°30′N and 10°10′N is one of the major estuaries of Kerala on the west coast of India (Fig. 1). The estuary receives a constant inflow of land runoff and industrial effluents. Vembanadu estuary acts as a sink for different trace metals. Point sources of pollution play an important role in the occurrence and distribution of heavy metals in the study area. The study area receives factory effluents mainly from Hindustan Newsprint Limited (HNL) from the major river Muvattupuzha. Accumulation and biomagnification of mercury in biota plays a significant role in the toxicity of this metal. Bed sediments and surface waters were analysed for the appraisal of the mercury content in the study area and other work underway.

Out of twelve sampling stations which were fixed, seven were from the Muvattupuzha river and five were from the Vembanadu lake (see Fig. 1). Surface water and bed sediment samples were collected using a precleaned plastic bucket during premonsoon (February, 2004) and monsoon (July, 2004) (Fig. 1). Nitric acid and potassium dichromate were added to water samples as preservatives for mercury analysis. The sediment is kept in low temperature until the analysis is carried out. Salinity and pH is determined by potable water quality analysers and percentage of organic matter is obtained by Walkey and Black method (Trivedi and Goel, 1986).

Mercury content in water and sediment is determined by mercury analyser, which is operated on Cold Vapour Atomic Absorption Spectrometry (USEPA, 1998). The sediment is dried and powdered and digested with aqua regia at 95°C in a water bath for 2 min, cooled, and added to milli-Q water and potassium permanganate (5%) solution and again kept in the water bath for 30 min. at 95°C. Cool and add hydroxylamine hydrochloride (12%) to reduce excess permanganate and make up to 50 ml (Anderson, 2000). The total

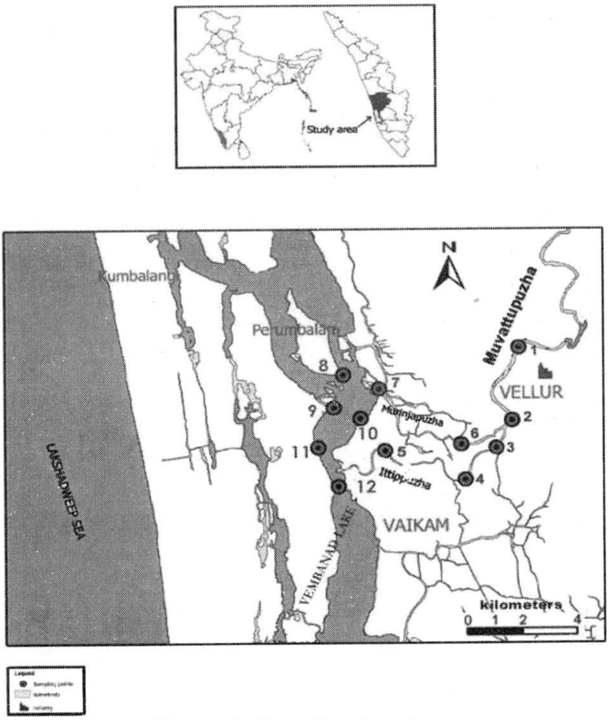

Figure 1: Sampling location.

mercury present in the water samples and digested solutions is reduced to elemental form by stannous chloride and measured by the absorbance at 253.7 nm with mercury analyser.

3. RESULT AND DISCUSSION

The analytical results are given in the Tables 1 and 2. pH varies from 6.9-7.5 and 6.55-6.86 in premonsoon and monsoon respectively. Slightly alkaline condition is prevailing in premonsoon than in the monsoon season. Salinity is varied during the two sampling months. In the monsoon season salinity recorded zero values in the river samples and it may be due to the large amount of fresh water discharges during southwest monsoon. In premonsoon season, stations 5 and 7, which are just near to the lake have shown some salinity (0.6 and 1.9 respectively). Salinity of lake samples varied from 8.5-14 and 0.2-0.3 during premonsoon and monsoon respectively. Salinity variation is due to influx of fresh water and the variation observed is taken as an index of estuarine mixing. Percentage of organic matter is ranged from 10.34-34.48 and 0.36-4.31 in premonsoon and monsoon months. High organic matter content is observed in the sediments taken during the premonsoon period. This may be due to the settlement of particulate matter in sediments.

Variation in mercury concentration in water samples during different seasons in the study area is shown in Figs 2 and 3. It is observed that in premonsoon season mercury is zero in all the water samples and observed little increased values during monsoon season, which might be attributed to flocculation. In the present context, the transport of particulate phase was high during monsoon, and because mercury is mainly associated with particulate matter and mercury is less soluble in water.

Table 1: Percentage of organic matter and total mercury in sediments

Station No.	*Organic Matter %*		*Total Mercury ng/g*	
	Premonsoon	*Monsoon*	*Premonsoon*	*Monsoon*
1	34.8	0.36	16.8	9.4
2	30.17	4.31	17.43	28.6
3	24.8	1.93	69.44	28.98
4	10.34	3.74	73.8	25.82
5	23.1	2.79	75.18	36.14
6	33.96	2.59	57.4	20.8
7	17.95	0.66	58.96	14.62
8	29.8	0.79	62.89	17.24
9	33.96	0.59	39.5	18.9
10	32.2	1.03	74.8	13.15
11	34.48	1.5	18.89	25.9
12	27.07	0.35	31.95	73.87

Table 2: Total mercury in water

Station	*pH*		*Salinity*		*Total Mercury ng/L*	
No.	*Pre-monsoon*	*Monsoon*	*Pre-monsoon*	*Monsoon*	*Pre-monsoon*	*Monsoon*
1	6.9	6.86	nil	nil	nil	nil
2	7.1	6.6	nil	nil	nil	1
3	6.95	6.69	nil	nil	nil	1
4	7.05	6.55	nil	nil	nil	nil
5	7.1	6.58	0.6	nil	nil	1
6	6.9	6.67	nil	nil	nil	1
7	7.04	6.63	1.9	nil	nil	nil
8	7.4	6.8	13.5	0.3	nil	1
9	7.5	6.78	8.9	0.2	nil	1
10	7.3	6.8	14	0.2	nil	2.5
11	7.3	6.76	8.5	0.3	nil	1
12	7.3	6.75	13.4	0.3	nil	nil

The analytical results of Hg in bulk sediment samples were given in Table 1. The total mercury in sediments ranged from 12.8-75.18 ng/g and 9.4-73.87 ng/g for the premonsoon and monsoon season. The present sampling and analysis were made with respect to point sources of mercury pollution. One sample is collected from the non-point sources and was noted as the

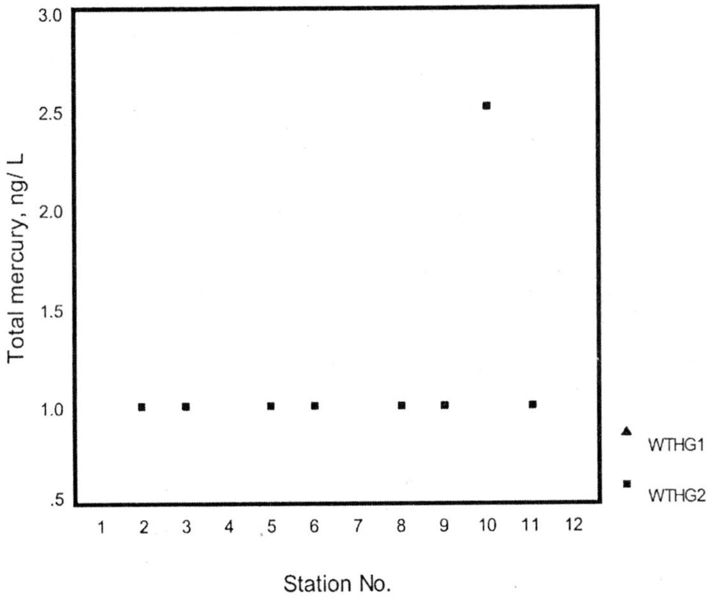

Figure 2: Total mercury in water (WTHG) (WTHG1– Premonsoon, WTHG2– Monsoon).

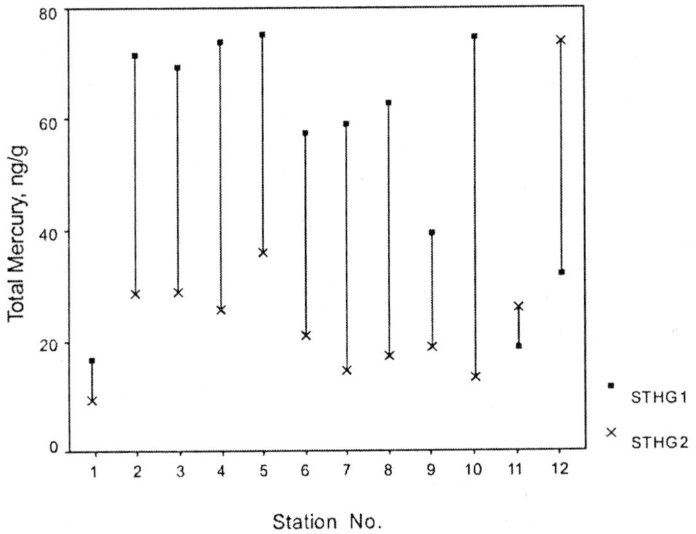

Figure 3: Total mercury in sediments (STHG).

control point (No. 1 in Fig. 1). The analytical results given showed that low concentration of mercury was observed in the control point. Generally, mercury content during premonsoon is high in all the stations except station no. 12.

There is possible correlation of total mercury content with respect to the percentage of organic matter and it is high in premonsoon. During monsoon period organic matter is discharged due to high influx of water and thus mercury content in sediment is low during monsoon period. The mercury concentrations obtained for the water and sediment of the Muvattupuzha river and Vembanadu lake are very less when compared to the concentrations obtained for other lakes and rivers (Jonasson and Boyle, 1971).

The high values given by the stations except control point show that the main source for this mercury is the pulp and paper industry situated on the banks of river Muvattupuzha. Salinity variations are not greatly affecting the river samples, but in the lake it is changing according to the tidal variations. These saline conditions can be the cause for the formation of organic forms by methylation, which is extremely toxic to aquatic food webs and also to human beings.

4. CONCLUSIONS

The concentration of total mercury in water is less than the permissible limit (0.001 mg/l) prescribed by BIS. The sediment mercury concentration is slightly higher than the normal terrestrial abundance (50 ppb) (Jonasson and Boyle, 1971). To avoid any future toxic hazards in the lake, further detailed investigations on the extent of mercury contamination in the Vembanadu lake and adjoining areas are necessary.

ACKNOWLEDGEMENTS

The author (MM) is thankful to State Committee on Science Technology and Environment, Kerala for providing financial assistance during the work and also to Dr. Baba, Director, Centre for Earth Science Studies and Dr. P.P. Ouseph, Head of the division for their valuable advise and suitable corrections.

REFERENCES

Anderson, K.A., 2000. Mercury analysis in environmental samples by cold vapour techniques in Encyclopedia of Analytical Chem: Instrumentation and applications. Meyers, R. A. (Ed), John Wiley & Sons Ltd, Chichester, pp. 2890-2903.

Cossa, D., Elbaz-Poulichet, F. and Nieto, J.M., 2001. Mercury in the Tinto-Odiel estuarine system (Gulf of Cadiz, Spain): Sources and dispersion. *Aquatic geochemistry*, **7:** 1-12.

Figueres, G., Martin, J.M., Meybeck, M. and Seyler, P., 1985. A comparative study of mercury contamination in the Tagus Estuary (Portugal) and major French estuaries (Gironde, Loire, Rhone). *Estuarine, Coastal and Shelf Science*, **20:** 183-203.

Hanson, A., 1971. Manmade sources of mercury. *In:* Proceedings of special symposium on mercury in man's environment. The Royal Society of Canada.

Jonasson, I.R. and Boyle, R.W., 1971. Geochemistry of mercury. *In:* Proceedings of special symposium on mercury in man's environment. The Royal Society of Canada.

Nriagu, J.O. (Ed), 1979. The biogeochemistry of mercury in the environment. Elsevier/North-Holland Biomedical Press, Amsterdam, New York, Oxford.

Porvari, V. and Verta, M., 2003. Total and methyl mercury concentration and fluxes from small boreal forest catchments in Finland. *Environmental Pollution*, **123:** 181-191.

Trivedy, R.K. and Goel, P.K., 1986. Chemical and biological methods for water pollution studies.

UNEP, 2002. Global Mercury Assessment.

USEPA, 1998. Determination of mercury in water by cold vapour atomic absorption spectrometry. Standard operating procedure - 515.

Sediment Dispersion in the Bay of Bengal

P.K. Mohanty, Y. Pradhan[1], S.R. Nayak[2], U.S. Panda and G.N. Mohapatra

Department of Marine Sciences, Berhampur University, Berhampur
pratap_mohanty@yahoo.com
[1]School of Earth, Ocean and Environmental Sciences (SEOES), University of Plymouth, Devon, PL4 8AA
[2]Indian National Centre for Ocean Information Services (INCOIS), Hyderabad

1. INTRODUCTION

The Bay of Bengal is about 2090 km long and 1610 km wide, bordered on the west by Sri Lanka and India, on the north by Bangladesh, and on the east by Myanmar (earlier Burma) and Thailand. The Andaman and Nicobar Islands separate it from the Andaman Sea, its eastern arm. The Bay of Bengal and the Andaman Sea are together defined as the oceanic area north of 5°N, bordered by the Indian subcontinent, Myanmar, Thailand, Malay Peninsula and Sumatra (Fig. 1). This unique semi-enclosed basin experiences seasonally reversing monsoons and depressions, severe cyclonic storms (SCS), and consequently receives a large amount of rainfall and river run-off in the tropics. It also encounters the largest seasonal sea level fluctuations (-40 cm to +54 cm) anywhere on the earth. An interesting characteristic of this area is its low saline surface water caused by large river run off from the Indian subcontinent and Myanmar. The circulation and hydrography of the Bay of Bengal is complex due to the interplay of semi-annually reversing monsoonal winds and the associated heat and freshwater fluxes. Apart from this, the inflow of warm high saline waters of the Arabian Sea, the Persian Gulf and the Red Sea origin and a number of synoptic disturbances (cyclones) originating during both pre-monsoon (May) and post-monsoon (October) period also affects the dispersal pattern in the Bay of Bengal. The impacts of the enormous discharge of riverine fresh water and sediments are least understood. However, the consequences could be severe, like changes in coastal morphology and the ecosystem since these rivers carry disposed sewage, industrial effluents, agricultural residues, etc. into the Bay of Bengal which contains higher concentration of Biochemical Oxygen Demand (BOD)

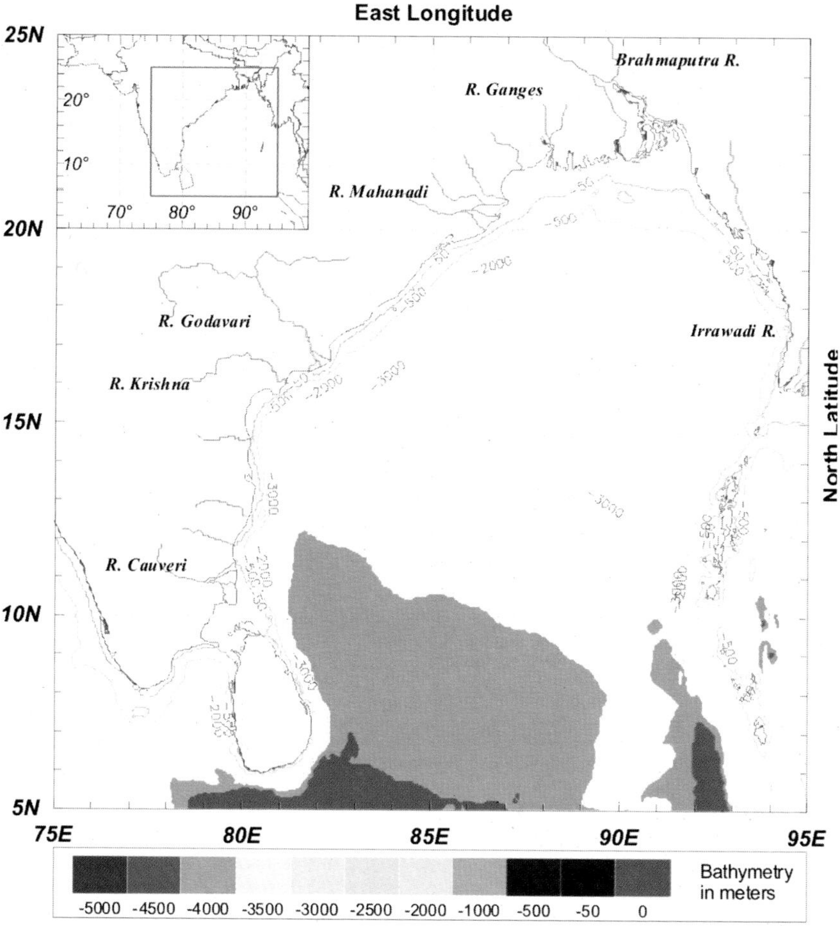

Figure 1: A glimpse of the northern Indian Ocean (inset) with the box showing the area of interest; major river systems, and bottom topography of the Bay of Bengal.

and faecal coliform. Due to the influence of water density and monsoon wind, the seasonal changes of the sea level in the Bay are remarkable and one of the highest in the world. It ranges from 166 cm at Khidirpur to 130 cm at Kolkata (Calcutta) and 118 cm at Chittagong. But towards the southwestern coast near Chennai (Madras) and Vishakhapatnam, the range is small compared to the northern and northeastern coasts of the Bay. The lowest variation of sea level at the southeastern coast of India is believed due to its geographical location at the edge of a comparatively deep sea. This is one of the probable reasons (accounting ~40%) for the differential sea levels between the Bay of Bengal and the Arabian Sea (Shankar and Shetye, 2001). Hence, it is very important to monitor and understand the fate of the freshwater discharge and the sediments. With the aid of satellite observation, it is now possible to revisit the source to sink pathways of this hefty fluvial discharge into the Bay. Remote sensing ocean colour data is a key parameter since it has the capability to estimate the suspended particulate matter empirically for light's interaction at selected wavelengths.

The Bay of Bengal is one of the largest fresh water and sediment input sites of the world ocean. The annual fresh water discharge into the Bay exceeds 1.5×10^{12} m^3 reducing the mean salinity by about 7‰ in the northernmost part. Fluxes of water are closely connected to the transport of sediment and dissolved constituents through river systems. The Bay receives about 2000 million tons of sediments annually contributed mainly through the Himalayan rivers—the Ganga and the Brahmaputra; Indian Peninsular rivers—the Mahanadi, the Godavari, the Krishna and the Kaveri, and the Irrawady and the Salween from the Myanmar.

2. CLIMATOLOGY OF RIVER DISCHARGE INTO THE BAY

The historical observations from the Global Runoff Data Centre (GRDC) database provide a unique opportunity to assess the volume discharges by the major river systems into the world ocean. The Global Monthly River Discharge Data Set contains monthly averaged discharge measurements for 1018 stations located throughout the world. The period of record varies widely from station to station with a mean of 21.5 years. The data are basically derived from the published UNESCO archives for river discharge, and checked against information obtained from the Global Runoff Centre in Koblenz, Germany through the U.S. National Geophysical Data Center in Boulder, Colorado.

Typically, river discharge is measured through the use of a rating curve that relates local water level height to discharge and generally river gauging is thought to have an accuracy of 5-10% (Vorosmarty et al., 1998).

Figures 2 and 3 show, respectively, the annual river discharge (from GRDC data) in different months and the GRDC stations corresponding to major rivers discharging into the Bay from the north and west. It is worth mentioning that measurement locations on the Ganges, the Brahmaputra and the Irrawady are about 300-400 km away from their mouths. These three rivers are perennial yet, the Ganges-Brahmaputra (hereafter G-B) system brings massive volumes during the summer monsoon. Fluxes form the Brahmaputra and the Irrawady remain almost flat during the entire summer monsoon whereas the Ganges, the Godavari and the Krishna show the peaks in August. Annual cycle of river discharge also gives a clear picture of negligible contribution by the Krishna-Godavari (hereafter, K-G) system during February through May, until the set of summer monsoon in June. The G-B system is still active and supplies more than 8000 m^3 s^{-1} during the winter monsoon.

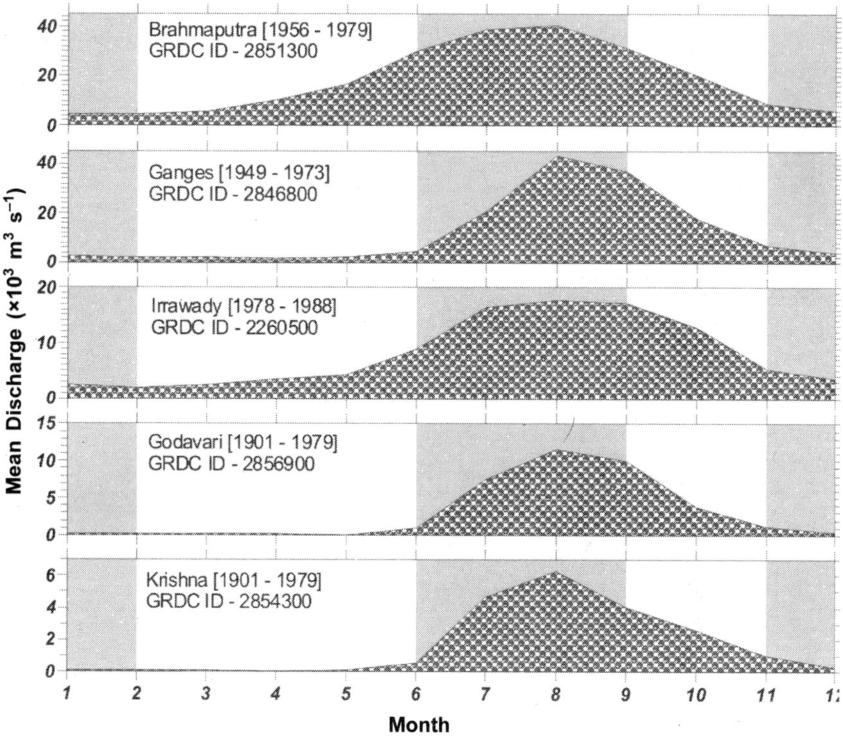

Figure 2: Monthly mean volume discharge into the Bay of Bengal at selected GRDC locations. Summer (SW) and winter (NE) monsoon months are highlighted with background grey shades.

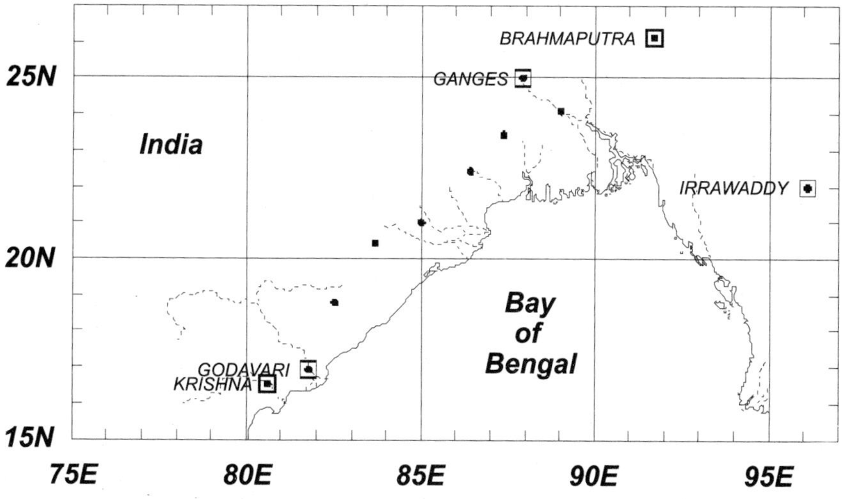

Figure 3: GRDC locations on different rivers (dots); Figure 2 has reference to the locations depicted as dots in squares.

The bounteous freshwater influx reduces the surface salinity considerably in the northern Bay. The hydrographic properties of the shelf region change drastically with seasons, especially in monsoons. Historically, it has been observed that surface salinities off Visakhapatnam of 29‰ and 34‰ in January and March, respectively, drop down to ~18‰ during the summer (SW) monsoon (Ganapathi and Murty, 1954). An SST gradient of ~1°C from north (26°C) to south (27°C) during winter is quite apparent in the Bay, which is basically instigated by the cold, dry north-easterlies aiding latent heat flux (evaporative cooling) and sensible heat flux (convection) from the sea surface in the northern Bay of Bengal. Another important aspect observed in the coastal Bay is the volume transfer, especially by the Himalayan rivers, in the northern sectors. The G-B system alone yields ~1000 × 10^6 tonnes/year of suspended sediment at a point around 200 km from the ocean in Bangladesh (Milliman and Meade, 1983), which is approximately 8% of the total sediment load reaching the global oceans (Milliman and Syvitski, 1992). This appears to be the highest suspended sediment load of any river system in the world and the large accumulation rate (665 × 10^6 t yr^{-1}) of sediments makes up the Bengal Delta and the 16.5 km thick Submarine Fan (Wasson, 2003), the largest deep sea fan in the world built up by turbidite deposits of the G-B origins (Kolla and Kidd, 1982; Emmel and Curray, 1984). Nath et al. (1989), from their geochemical analyses of deep-sea sediments in the Central Indian Basin, confirmed that the sediments from the G-B source are transported even south of the equator up to 8°S, covering a large distance of over 3000 km from the river mouth. Goodbred and Kuehl (1989) estimated the deposition as ~21% in the sub-aqueous delta and ~29% in the fan, of the combined (G-B) river transport (980 × 10^6 t yr^{-1}).

Table 1: Sediment yield and volume run-off by the major riverine systems to the Bay of Bengal

River System	*Basin area (× 10^3 km^2)*	*Run-off (× 10^6 m^3 yr^{-1})*	*Sediment load (× 10^6 tonnes yr^{-1})*
Ganga	750.0	493.0	329.0
Brahmaputra	580.0	510.0	597.0
Irrawady	430.0	422.0	265.0
Godavari	313.0	92.2	170.0
Krishna	251.4	32.4	4.0
Mahanadi	41.0	54.5	15.7
Brahmani	28.2	16.3	20.4
Cauvery	66.3	21.5	1.5

Source: Subramanian (1993)

Table 1 briefly summarises the sediment yield and volume run-off to the Bay by these rivers (River Brahmani, apart from the major seven rivers, is included in the list here since it has a significant contribution to the sediment load). The huge volume of freshwater influx (Table 1) together with monsoon winds strongly influences the circulation, stratification, productivity and sedimentation pattern in the Bay of Bengal. Vertical stratification, in the Bay, due to temperature variation holds for greater depths (>100 m); however, salinity dominates this phenomenon in shallow waters which is not very commonly observed elsewhere. The effect of freshwater discharge on particle fluxes and fate of terrigenous material discharge are demonstrated respectively by Ittekkot et al. (1991) and Ramaswamy et al. (1997). The dispersal of sediments within the province is affected primarily by surface oceanic circulation and by bottom and turbidity currents (Kolla et al., 1976). Even though the seasonally reversing East India Coastal Current (EICC), which is about 200 m deep and 100 km wide (Wyrtki, 1973), plays an important role in the sediment transport along the central and south-central part of the east coast of India, large freshwater influx from the G-B system during the SW monsoon overpowers other processes giving rise to equator-ward freshwater plumes against the prevailing local winds (Shetye et al., 1991).

Table 2: Seasonal runoff volumes of major rivers

River system	*Annual mean* (m^3s^{-1})	*NE (DJF) monsoon ratio*	*SW (JJA) monsoon ratio*	*Total Run-off volume* ($\times 10^9$ m^3)	
				NEM	*SWM*
Mahaweli	226	0.42997	0.16267	3.0645	1.1549
Godavari	3180	0.02198	0.55407	2.2043	55.5646
Krishna	1730	0.02089	0.61949	1.1397	33.7977
Mahanadi	1710	0.02109	0.59933	1.1373	32.3198
Pennar	95	0.20271	0.13814	0.6073	0.4193
Damodar	329	0.03059	0.56124	0.3174	5.8231
Ganga	11892	0.06532	0.49260	24.4967	184.7363
Brahmaputra	16186	0.05949	0.49552	30.3672	252.9340
Cauvery	664	0.20271	0.13814	4.2447	2.8926
Irrawady	13018	0.04863	0.52915	19.9660	217.2345
Salween[1]	5421	0.04863	0.52915	8.3143	90.4615
Meghna[2]	4215	0.05949	0.95520	7.9079	65.8666
([1]in Myanmar and [2]in Bangladesh)			Total	103.7673	943.2049

Source: Rao and Murty (1992)

The results of copious, pulsed freshwater and terrigenous discharge during the SW monsoon (Ittekkot et al., 1985; Ittekkot, 1993) are observed as a fair reduction in the surface salinity (more than 7‰ in the northern Bay) over the entire Bay (LaViolette, 1967; Wyrtki, 1973; Murty et al., 1990). Most of the freshwater influx to the Bay occurs during the southwest monsoon. As reported by Ittekkot et al. (1985), more than 80% of the annual water discharge in the Ganges occurs between July and November, during which more than 80% of the annual sediment discharge is accounted with suspended matter concentrations up to 1250 mg l^{-1}. Typical figures of volume runoff during the two monsoons are depicted in Table 2 (Rao and Murty, 1992). It is obvious that except for some southern peninsular rivers (Cauvery, Pennar and Mahaweli), all other rivers discharge maximum volume during the SW monsoon.

Figure 4 shows the inter-annual variability of volume discharge in a 20-year (1960-1980) time-series. The seasonal cycle is strongly reflected in all small seasonal and large perennial rivers. During the peak months, the volume discharge varies from as low as 3×10^4 m^3s^{-1} to as high as 6×10^4 m^3s^{-1} by both the Ganges and the Brahmaputra rivers. The amount of flux also decreases gradually towards the mouth as the total volume is distributed among the tributaries. The upper two panels (a and b) of Fig. 4 show the volume discharge at Panda and Bahadurabad in the Brahmaputra. Figures 4c and d show the volume discharge at Paksay and Farakka in the Ganges respectively. The volume discharges by four peninsular rivers are shown in Figs. 4e-h. The massive flux is contributed from the Himalayan source apart from the seasonal contribution by the peninsular rivers. The off season discharge by the G-B is almost equivalent to (sometimes greater than) the volume discharge by any single peninsular river (Fig. 4).

The summer monsoon rainfall index (Fig. 5) is an important parameter which controls the river discharge during the period. It is clear, however, from Figs 4 and 5 that the one-to-one correlation between monsoonal rainfall and volume discharge is predominant for the peninsular rivers. The Himalayan rivers often show discrepancies from the summer monsoon rainfall leaving a complicated trace to understand the complex seasonal dynamics at the mouth. Hence it is important to find out the fluvial pathways of the huge volume influx by the G-B system.

3. CIRCULATION IN THE BAY AND SEDIMENT DISPERSION

The circulation in the Bay is a complex phenomenon because of the seasonally reversing monsoon wind forcing coupled with the large volume of fresh water discharge from various rivers. Retrospective studies on the circulation

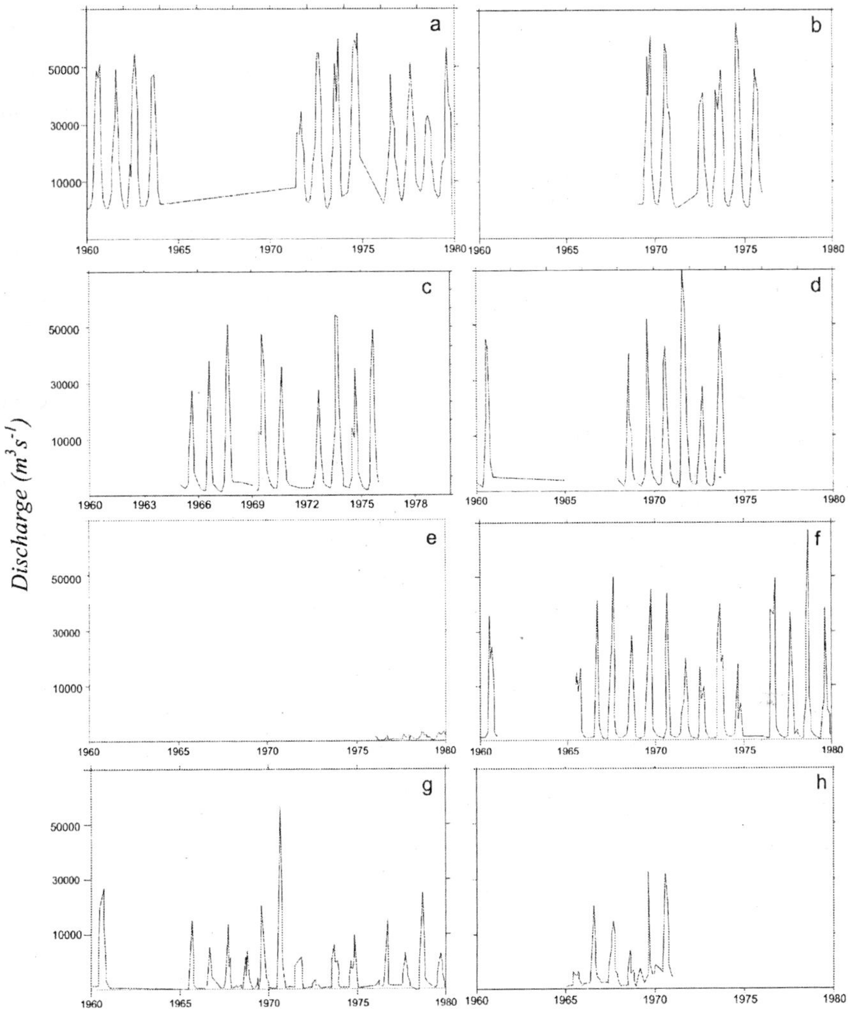

Figure 4: Time series of historical measurements (1960-1980) of net volume discharge by the Himalayan and Peninsular rivers. For River Bramhaputra at (a) Panda (91.70E/26.13N) and (b) Bahadurabad (86.66E/25.18N); for River Ganga at (c) Paksay (89.03E/24.08N) and (d) Farakka (87.92E/24.83N); (e) for River Cavery (78.83E/10.83N); (f) for River Godavari (81.78E/16.92N); (g) for River Krishna (80.62E/16.52N) and (h) for River Mahanadi (83.67E/20.42N).

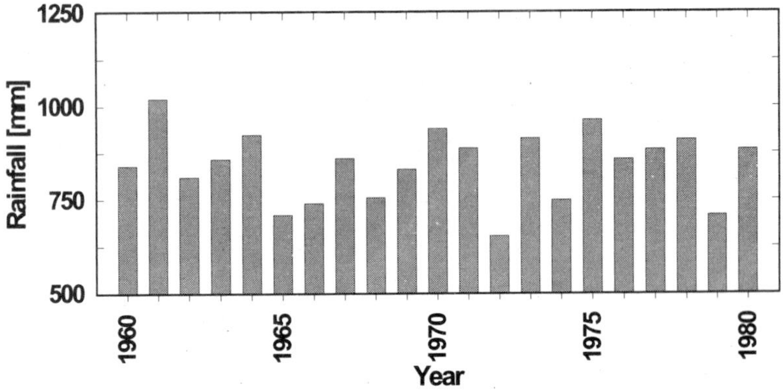

Figure 5: Summer monsoon rainfall (June-September) indices over India for the period 1960-1980. (Data source: IRI/LDEO Climate data library, downloadable from http://iridl.ldeo.columbia.edu/SOURCES/.Indices/.india).

of the Bay of Bengal (La Fond and La Fond, 1968; Varadachari et al., 1968; Duing, 1970; Wrytki, 1971; Rao, 1977; Rao and Sastry, 1981) are based on the hydrographic data, limited both in space and time, who suggested the circulation to be predominantly anti-cyclonic during pre-monsoon (March-May) and cyclonic during post-monsoon (October-November) (La Fond and La Fond, 1968). Based on the *International Indian Ocean Expedition* (IIOE) data, Duing (1970) suggested that the circulation in the bay is anti-cyclonic throughout the year. In an experiment on the circulation and geostrophic transport in the Bay of Bengal, Rao and Murty (1992) observed the circulation in the Bay as primarily cyclonic surrounded by cyclonic and anti-cyclonic gyres during the beginning of both the monsoons. They also mentioned that the circulation during NE monsoon consists multiple-cellular cyclonic and anti-cyclonic gyres extending to deeper depths. During the northern hemisphere winter (boreal winter), the main features of the ocean currents include large-scale anticyclonic flow in the Bay of Bengal surface waters (Potemra, 1991). This gyre decays into eddies in spring and then transition into a weaker, cyclonic gyre by late summer.

In modern oceanography, the general circulation in the Bay of Bengal is characterised by anti-cyclonic flow during most months and strong cyclonic flow during November. Currents are weak and variable in January. In the west, the *East Indian Current* (EIC) strengthens as the NE Monsoon becomes stronger, exceeding 0.5 ms^{-1} in March (Fig. 6) and remaining strong (0.7-1.0 ms^{-1}) until May/June (Tomczak and Godfrey, 2001). Throughout this time the current runs into the wind, apparently as an extension of the *North Equatorial Current* (NEC). During the SW monsoon season, currents in the entire Bay are weak and variable again. The highest velocities (around 0.5 ms^{-1}) are found in the *East Indian Current* and the flow along the eastern coast rarely exceeds 0.2 ms^{-1} but is often directed into the wind. An indication of a current reversal in the west is seen in September (Fig. 6). Currents are consistently southwestward and strong (0.5 ms^{-1} and more) north of 15°N and close to the shelf southwestward flow prevails.

Complete reversal of the *East Indian Current* into the *East Indian Winter Jet* (EIWJ) is not achieved until late October, when water from the Equatorial Jet enters the Bay in the east and a cyclonic circulation is established. The *East Indian Winter Jet* (EIWJ) is a powerful western boundary current with velocities consistently above 1.0 ms^{-1}. It follows the topography south of Sri Lanka and feeds its water into the Arabian Sea. Very little exchange occurs with the *Equatorial Jet* (EJ) south of Sri Lanka; currents in the separation zone between the two jets (near 3°N) are weak and variable. The *East Indian Winter Jet* (EIWJ) fades away from the north in late December, its southern part merging with the developing *North Equatorial Current* (NEC).

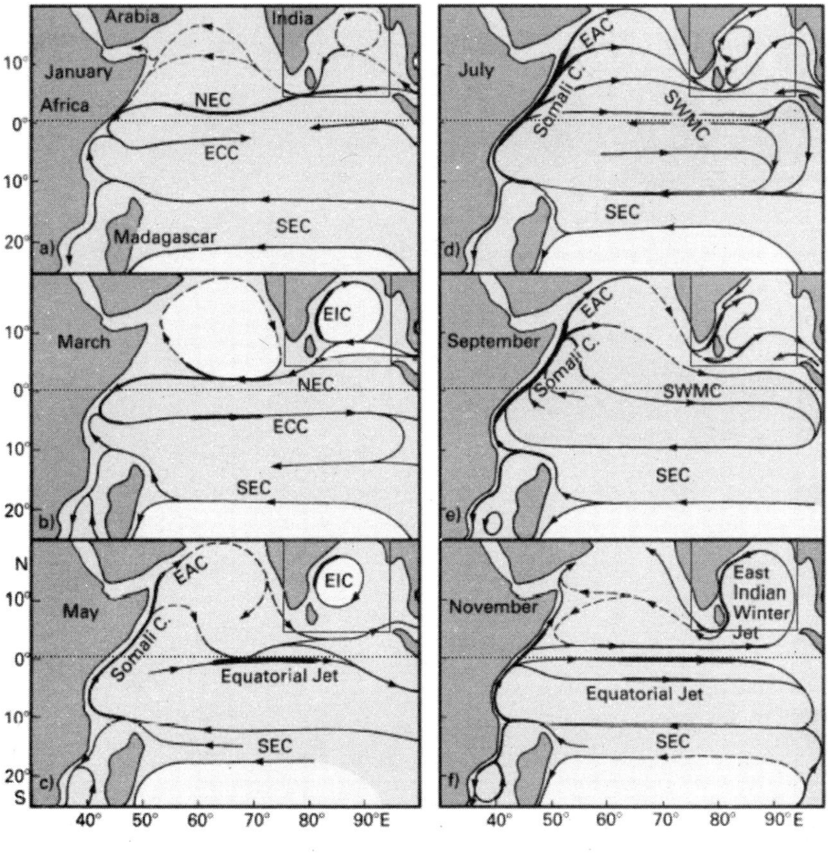

Figure 6: Surface Currents in the Indian Ocean. Notice the seasonal changes in the Bay of Bengal region (highlighted in square box). Adapted from Tomczak and Godfrey (2001) after Cutler and Swallow (1984).

3.1 Seasonal Circulation

The Bay of Bengal is distinguished by strong near-surface stratification. Hence, it is anticipated that the physical properties of the upper layers, for example, surface currents and temperature, would exhibit large variability in the spatial domain. According to Potemra et al. (1991), the seasonal circulation in the Bay of Bengal can be separated into four stages:

(a) A large anti-cyclonic gyre across the whole Bay during December until March;
(b) This is followed by two counter-rotating flows, anticyclonic on the western side and cyclonic on the eastern side, producing northward currents along both coasts and southward flow down the middle;
(c) The previous pattern persists until early summer (April-June), when the anti-cyclonic flow extends across the whole Bay, and in the SW monsoon months (July and August) is characterised by counter-clockwise flow; and
(d) Finally, in the autumn, two rotating flows develop again, with southward current along both coasts and northward flow in the centre.

Varkey et al. (1996) explained the seasonal circulation by a three-gyre circulation pattern (G1, G2, G3) in a schematic way for the two monsoons – summer or SW monsoon and winter or NE monsoon (Fig. 7).

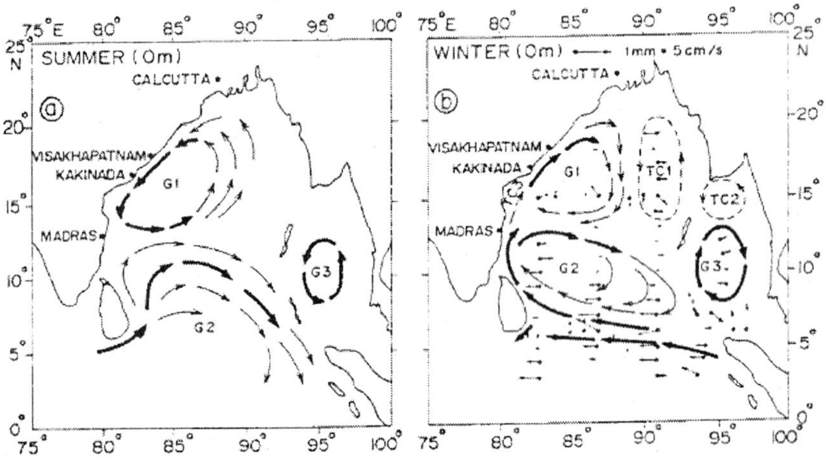

Figure 7: Schematic of the seasonal surface circulation in the Bay of Bengal (After Varkey et al., 1996).

The northern gyre (G1), between the western boundary and 13°N/89°E, is clockwise during winter and anticlockwise in summer. They observed a reversal of G1 at depths greater than 500 m during winter (not illustrated here), which needs further explanation. A southern gyre (G2) in the area south of 13°N is clockwise during both winter and summer. The gyre in the

Andaman Sea (G3) has clockwise and anticlockwise flows in winter and summer, respectively. The steadiness and strength of these gyres and other currents in the Bay seems to depend on the development and shift of the *North Equatorial Current* (NEC) and the *Indian Monsoon Current* (IMC). In numerical experiments Yu et al. (1991) showed that the seasonal reversal of gyre G1 in the northwestern Bay is caused partly by remote forcing due to monsoon winds in the equatorial Indian Ocean. They, using a reduced-gravity model, found that the long *Rossby waves* excited by the remotely forced *Kelvin waves* contribute substantially to the variability of the local circulation.

Cutler and Swallow (1984) compiled (averaged) the historical surface current data collected by the British Meteorological Office for over a century (1854-1982) with 1° × 1° space and 10-day time grid, which confers a precise awareness about the seasonal cycle of the near-surface circulation in the Bay. At the outset of the NE monsoon, the Bay has a basin-wide cyclonic circulation. Schott et al. (1994), from the shipboard current measurements and from moored instruments during 1991-92, observed a reversal and reduction of the near-coastal transport, which they suspected to be a result of the *Kelvin waves* from the Bay of Bengal.

3.2 The Monsoon Currents

The monsoon currents are essentially Ekman drifts forced by the monsoon winds, the geostrophic contribution to these flows being negligible. The monsoon current varies round the year. In January (peak NE monsoon), the westward flowing Indian Monsoon Current (IMC) south of Sri Lanka is supplied from the east (south of 8°N), which weakens during the inter-monsoon (March) period but still fed by the well-developed southern gyre. The reverse flow (eastward) of the Monsoon Current is observed throughout the SW monsoon period following the onset with a slight break during July, picking up in speed again in August. Shankar et al. (2002) demonstrated the monsoon currents as a trans-basin phenomenon in the northern Indian Ocean using an Oceanic General Circulation Model (OGCM). They showed that the westward flowing *Winter Monsoon Current* (WMC) develops south of Sri Lanka in November that is initially fed by the equatorward East India Coastal Current (EICC).

The WMC in the Bay appears later in the following months (Fig. 8). The eastward flowing *Summer Monsoon Current* (SMC) continues to flow from the Arabian Sea passing through the Lakshadweep low and splits into two branches after crossing the Sri Lankan coast—one into the Bay of Bengal and the other in the eastward direction. Net transport due to the shallow monsoon currents is attributed to both Ekman drift and geostrophic flow. In the Bay of Bengal, the monsoon currents are generally forced by Ekman pumping and by the winds in the equatorial Indian Ocean. However, the hydrographic data show that the monsoon currents are not found in the same

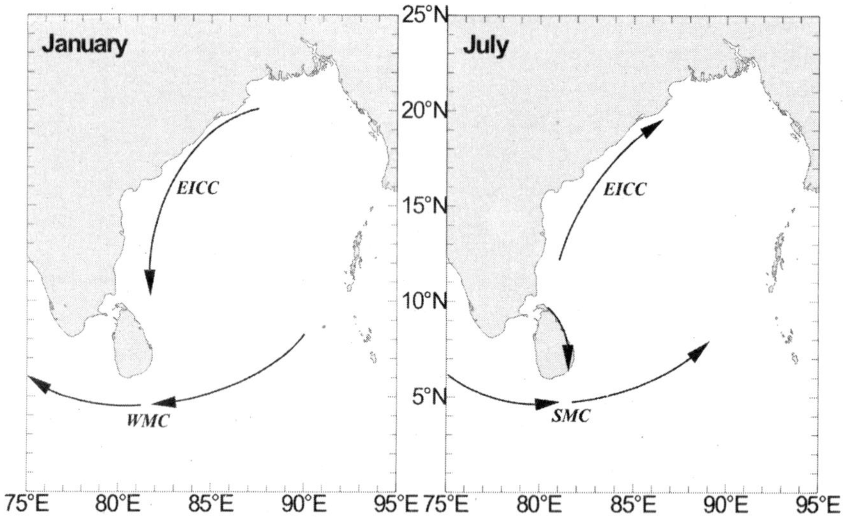

Figure 8: Schematics of the current systems in the Bay of Bengal during January (winter monsoon) and July (summer monsoon) modified after Shankar et al., 2002. Legends: WMC – Winter Monsoon Current, SMC – Summer Monsoon Current, EICC – East India Coastal Current, SL – Sri Lanka.

location during a season or across different years; for example, Vinayachandran et al. (1999) showed that the SMC in the Bay of Bengal intensifies and shifts westward as the summer monsoon progresses.

3.3 East India Coastal Current (EICC)

During last few years a number of studies have helped to identify the principal mechanism that drives the general circulation in the Bay of Bengal (Potemra et al., 1991; Yu et al., 1991; McCreary et al., 1993) and the East India Coastal Current (EICC) as an extension of the western boundary currents (Shankar et al., 1996; McCreary et al., 1996). Western Boundary Currents (WBCs) persist in response to the large-scale zonal wind systems along with the combined effect of the curvature and rotation of the Earth. The WBCs in the major oceans have always fascinated oceanographers by virtue of their well-developed and intense flow patterns and they have potential for generating warm/cold core eddies which are an integral part of the general circulation. Though numerous studies have been made on WBCs viz. the Gulf Stream and the Brazilian Current in the Atlantic, the Kuroshio and East Australian Current in the Pacific, the Aghulas Current in the Southern Indian Ocean, and the Somali Current in the western Indian Ocean, such studies are meagre in the Bay of Bengal where the surface flow reverses seasonally. However, based on the size of the oceanic basin, the Bay of Bengal is a possible location for WBCs. But the currents in this basin are variable due to the biannual cycle of the monsoon winds. Like the Somali Current in the Arabian Sea, the currents along the east coast of India reverse their directions twice a year, flowing northeastward from February until September with a strong peak in March-April and south-eastward from October to January with strongest flow in November. Recently the western boundary current in the Bay of Bengal has been named as the East India Coastal Current (EICC). The current along the east coast of India is hereafter referred to as EICC. It is observed that the EICC changes its direction twice a year (see Fig. 8) flowing north-eastward from February until September with a strong peak in March-April and southwestward from October to January with strongest flow in November (Shankar et al., 1996; McCreary et al., 1996). The current extends to a depth of 200 m and has a transport of ~10 Sv. The recent study using NOAA AVHRR imagery by Ratna Reddy et al. (1995) showed the current about 900 km in length, usually lying close to the coast but occasionally shifting offshore (Shetye et al., 1993), with an average speed of 30-55 cm s^{-1}. The SST within the boundary current is reported as ~27°C, and the temperature on either side of this current is lower (~26°C). This western boundary current is also a part of an eddy field like the Somali Current in the SW monsoon (Shetye et al., 1993). Yu et al. (1991) proposed a bifurcation of the WBC along the East Indian coast at ~12°N with a warmer poleward and a colder equatorward current. There are some hydrographic studies (e.g. Shetye et al., 1993; Shetye et al., 1996; Sanilkumar et al., 1997) and numerical studies (McCreary et al., 1996; Shankar et al., 1996) further describing the EICC and its driving mechanism. Studies made in the recent past (Yu et al., 1991; Potemra et al., 1991; McCreary et al., 1993; Shankar et al., 1996; Shetye et al., 1996; McCreary et al., 1996) have suggested four different principal mechanisms that drive the EICC:

(a) Interior Ekman pumping over the Bay;
(b) Local alongshore winds adjacent to the east coast of India and Sri Lanka;
(c) Remote alongshore winds adjacent to the northern and eastern boundaries of the Bay;
(d) Remotely forced signals that propagate into the Bay from the equator.

4. COASTAL PROCESSES AND SEDIMENT DISPERSION

4.1 Tides and Fronts

The processes in the coastal waters are greatly complicated by factors peculiar to the coastal zone, viz., the shallowness, the presence of tidal currents, river run off, and the barrier to advection posed by the coastline itself. In the presence of strong tidal currents that create turbulence in shallow waters, tidally induced mixing may extend all the way to the surface. This, in conjunction with the flow of freshwater from the land, makes the dynamics more complicated in the Bay. The less haline (~0 psu) and much lighter freshwater by lying on the top of the dense seawater creates a stratification that can be independent of temperature difference of the layers, and forms the buoyancy-driven currents. So, one way of trying to understand the complex relationship existing between physical and biological processes in coastal waters is to view freshwater run-off as a mechanism tending towards greater stratification while wind-driven and tidal currents are mechanisms tending to cause turbulence in the water column and to breakdown stratification. In the shallower regions (e.g., head of the Bay), the effects of tidal mixing lead to the formation of distinct fronts of different space and time scales which often overlap with the shelf-break fronts. These shallow-water fronts have biological importance since they account for high densities of phytoplankton concentrations. Since the east coast of India has a very narrow and stiff continental self in the central and southern sectors, local plume fronts are observed as a consequence of seasonal reverine discharge. It is worth mentioning here that the fronts can also be created by large turbulent eddies, or by the coastal upwelling.

Both semi-diurnal (M_2, principal lunar) and diurnal (K_1, principal lunar/ solar) constituents contribute to the tides along the coasts surrounding the Bay of Bengal (Fig. 9). Tides in the Bay are mixed semi-diurnal in nature (two high and two low tides every lunar day, i.e. 24 hours 50 minutes). The highest tide is seen where the influence of bottom relief and the configuration of the coast are prominent, i.e. in shallow water and in the northern Bay and estuary. The mean height of tidal waves near the coast of Sri Lanka is around

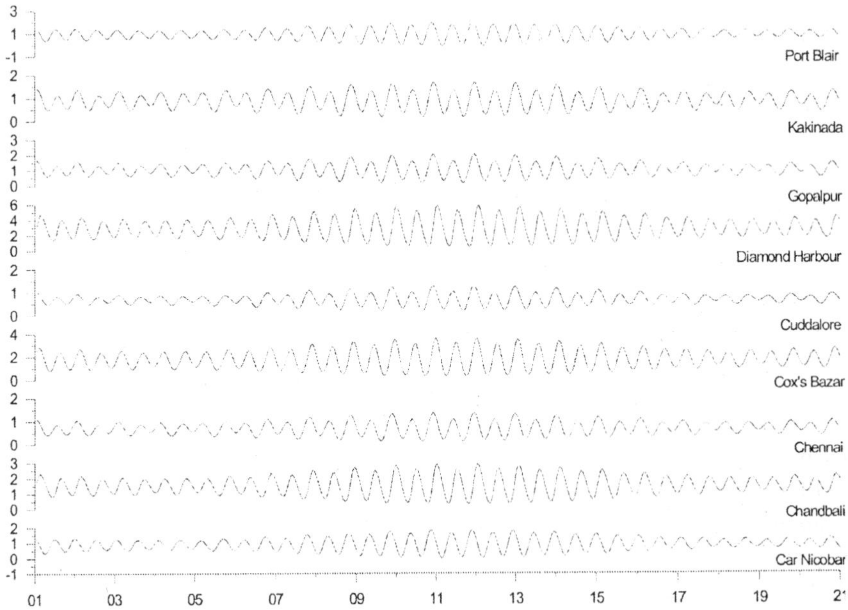

Figure 9: Typical tide curves over a period of three weeks at selected ports along the coasts surrounding the Bay of Bengal.

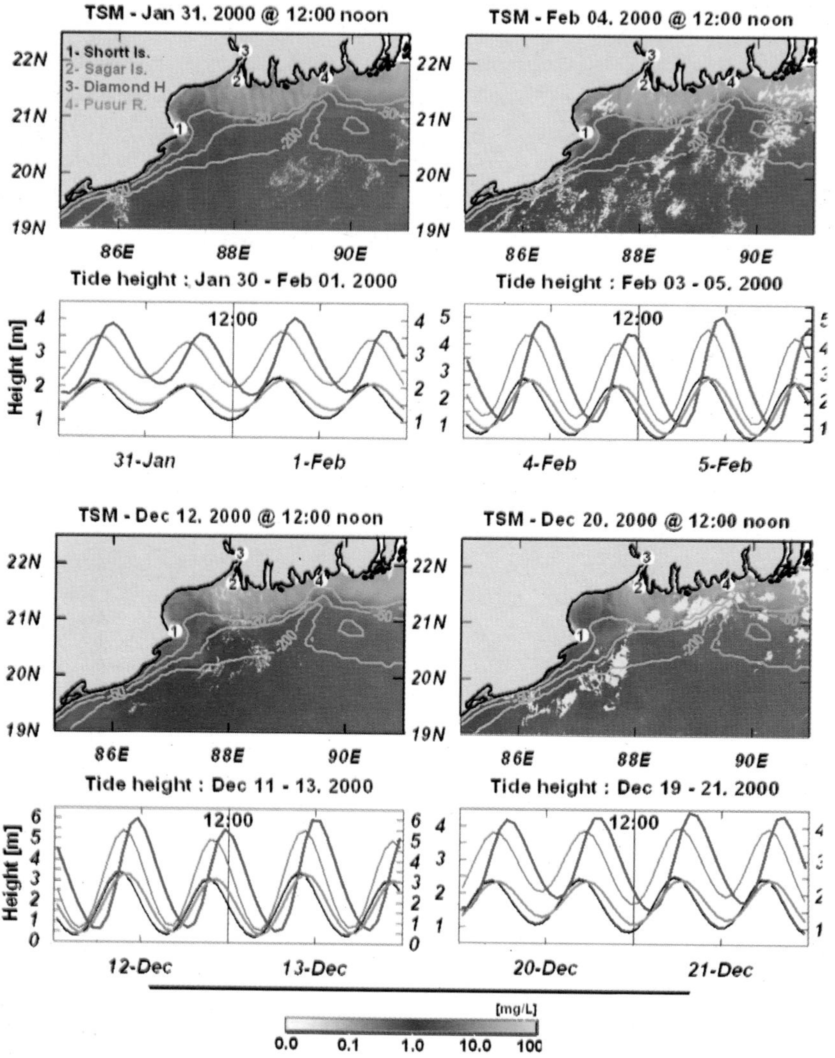

Figure 10: Total Suspended Matter (TSM in mg ℓ^{-1}) derived from IRS-P4 Ocean Colour Monitor (OCM) data. Bathymetry contours for 20 m, 50 m and 200 m are overlaid on the TSM maps. Tide curves at four different locations (1–Shortt Island, 2–Sagar Island, 3–Diamond Harbour and 4–Pusur river) for corresponding dates are plotted below the TSM maps.

0.7 m whereas it is 4.71 m near the deltaic coast of the Ganges. In the Bay of Bengal tidal currents specially develop in the mouths of the rivers, for example the Hooghly. The currents associated with the semi-diurnal tides play significant role in material transport and distribution along the coasts in the Bay of Bengal. Figure 10 shows the surface distribution of TSM, during low-tide and slack period between high and low tides, at four selected locations in winter 2000. Tide heights range at Shortt Island and Pusur River mouth are almost half of the range at Sagar Island and Diamond Harbour.

Frequent excursion of suspended matters keeps the shelf zone (below 20 metre depth) ever turbid with TSM values exceeding 50 mg ℓ^{-1}. The TSM snaps on 31st January and 20th December were taken during near low-tide conditions (when the magnitude of tidal stream currents are minimum), which reveal the shoals and their elongated patterns more clearly as the water level recedes although the exposure of these shoals are partially hindered by clouds on 20th December. On the other hand, the TSM images on 4th February and 12th December were taken during the transition between high and low tides when the tidal current is at the peak and offshoreward. Sediment plumes near station 1 are observed to be stretched towards the 50 m depth isoline. The plumes in the north which were strong within the 20 m depth contours have now stretched up to 50 m. This observation sets two ideas regarding sediment transport across the northern Bay – (1) the elongated NE-SW and NW-SE shoals across the *Swatch of no Ground* essentially indicate the tunnelling of the materials and (2) a branch of the plume west of 88°E also moves southwestward along the 20-50 m bathymetry track which often hugs the coast.

4.2 Surface Current: Role of Ocean Colour Data

The patterns of ocean colour on sequential images can be used as tracers, to a better extent than SST, to measure displacements of surface waters. Garcia and Robinson (1989) first showed the use of ocean colour data to extract sea surface velocities in shallow seas. The objective method is used to estimate the surface currents from OCM derived TSM maps. The method is based on matching suspended sediment dispersal patterns in two sequential time lapsed images. The movement of the pattern can be calculated knowing the

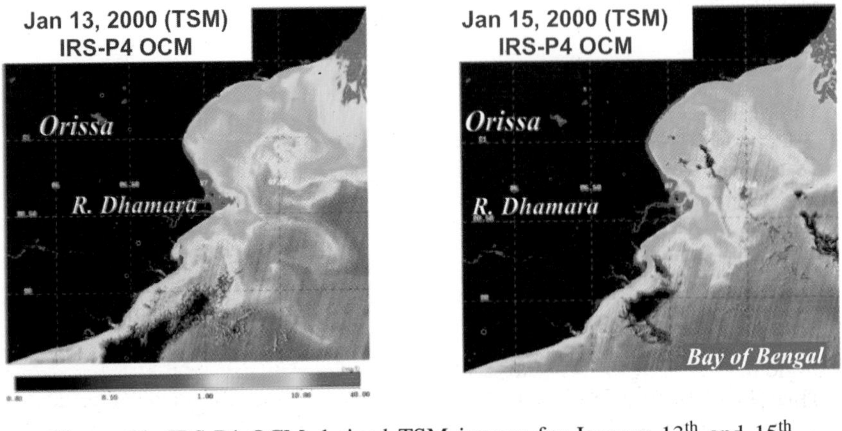

Figure 11: IRS-P4 OCM derived TSM images for January 13th and 15th, 2000 near Dhamara estuary, north Bay of Bengal.

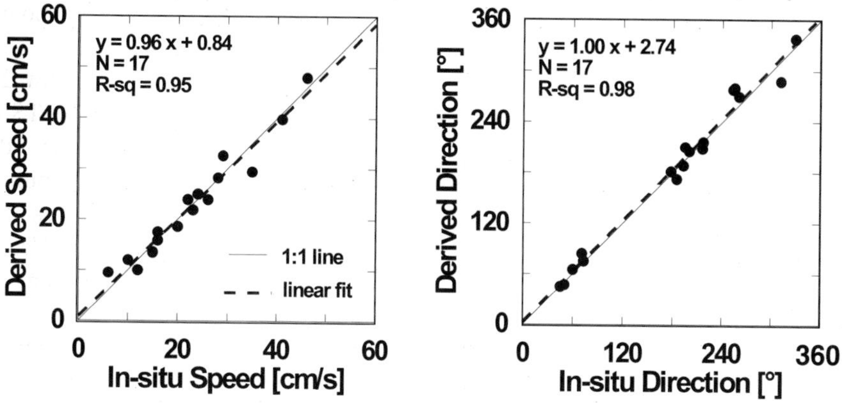

Figure 12: Scatter plots of in-situ vs derived surface current speeds (left panel) and directions (right panel).

displacement of windows required to match patterns in successive images. Figure 11 shows the TSM images of 13th and 15th January 2000, on which the Maximum Cross Correlation (MCC) scheme was applied and the currents were retrieved. The scatter plots between retrieved and measured current speeds and directions (only during local noon) are shown in Fig. 12 that essentially explains the validity of this technique. The study also demonstrates the feasibility of ocean colour data to understand the transport velocity and pathways of optically active near-surface tracers.

4.3 Effects of Oceanic Eddies in Coastal Waters

To improve the observational strategy in understanding the effects of medium-large eddies on coastal circulation, satellite data from different disciplines are used in this section.

Figure 13 shows the net surface flow vectors (estimated from T/P and QS data) on OCM derived TSM map. The surface currents are generated from the weekly data. The TSM maps give reliable pictures on the surface advection in the nearshore waters. In the top panel of Fig. 13, an elliptical anti-cyclonic eddy of about 400 km diameter along its major axis is seen near 86°E/18.5°N during the first week of March. Away from the northern-most shelf (where tides control the flow), the sediment plumes off the Mahanadi delta, follows the surface currents along the northern periphery of the eddy as far as 100-120 km from the mouth. Similarly, in the bottom panel three distinct plumes are visible at the Godavari, the Krishna and Pennar river mouths. Interestingly, the plumes off the Godavari and the Krishna moves at right angle to each other (southward off the Godavari and eastward off the Krishna). This is true since the coastal circulation during this period was affected by an eddy triplet. A large cyclonic eddy (centred near 86°E/16°N), sandwiched between the two anti-cyclonic eddies (off the Chilika lagoon and Pennar

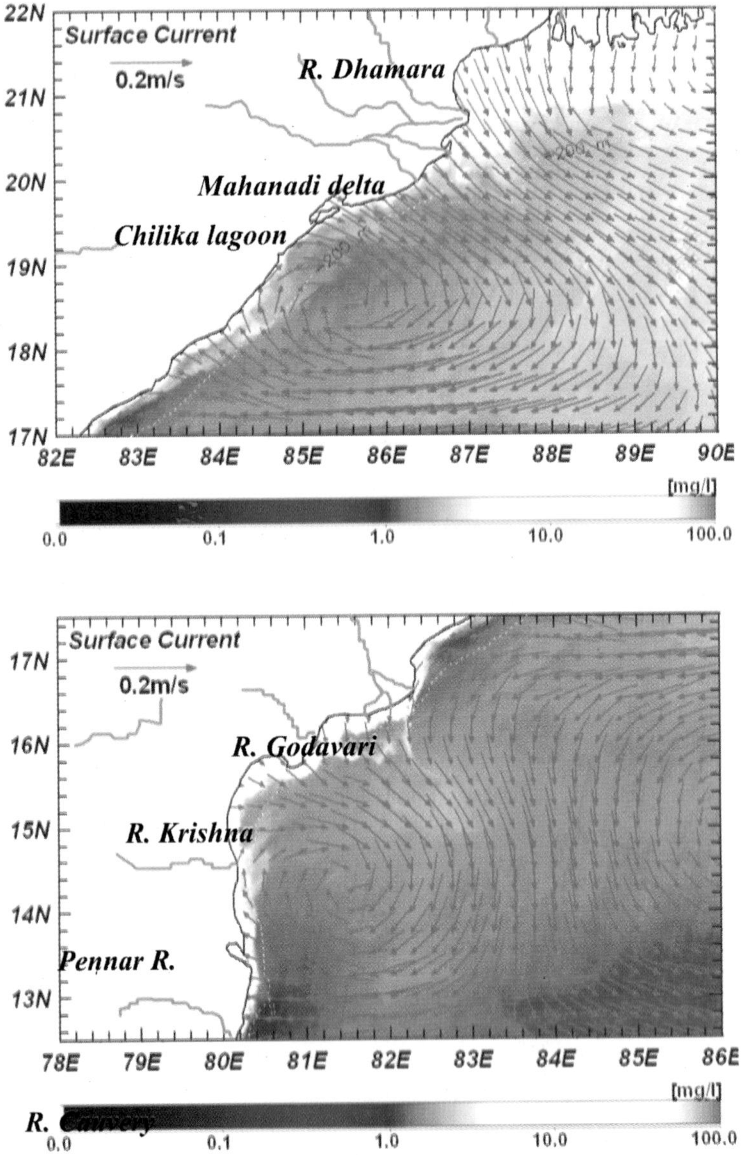

Figure 13: Net surface flow (in $m.s^{-1}$) estimated from T/P-QS data for March week#1, 2000 overlaid on OCM TSM (in $mg.l^{-1}$) maps of March 01, 2000.

river) seems to drive the Godavari plume towards south whereas the southern eddy moves the plumes off the Pennar, the Krishna northward and eastward, respectively. Hence it would be quite valuable to understand the dispersal pathways through the estimation of surface currents using microwave data during the cloudy seasons.

5. DISPERSAL PATHWAYS

It was reported in the past that sediments from the G-B system are transported into the southern hemisphere (Nath et al., 1989) as far as 8°S, covering a distance of more than 3000 km. The *Swatch of no Ground*, also known as the *Ganga Trough*, has a comparatively flat floor 5-7 km wide and walls of about 12° inclination (Fig. 14). At the edge of the shelf, depths in the trough are about 1200 m. The *Swatch of no Ground* has a seaward continuation for more than 2000 km down the Bay of Bengal in the form of fan valleys with levees. The sandbars and ridges near the mouth of the G-B delta pointing toward the *Swatch of no Ground* suggest that sediments are tunnelled through this trough into the deeper part of the Bay of Bengal. Thus, the *Swatch of no Ground* is a potential region to transport the sediments beyond the continental shelf into the deep Bay. Some of the shoals and sand ridges

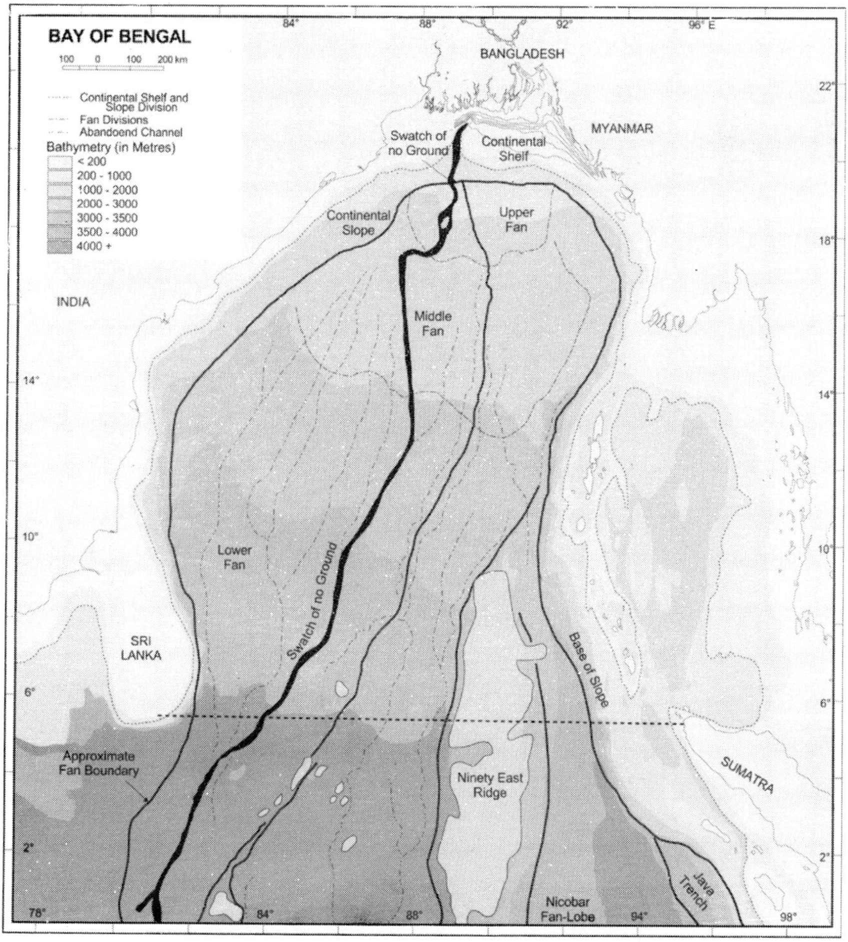

Figure 14: The Bay of Bengal and the environs showing the different regions of the Bengal Fan and the Swatch of no Ground.

present on this part of the continental shelf show an elongation pattern pointed towards the *Swatch of no Ground.* Bengal Deep Sea Fan, also known as Bengal Fan, is the world's largest submarine fan. The Bengal Fan has been built up principally by turbidite deposits of the G-B origin (Emmel and Curray, 1984). Together with its eastern lobe, the Nicobar fan, it covers an area of 3106 sq km with approximate length and width of 2800-3000 km and 830-1430 km, respectively and a thickness of more than 16 km beneath the northern Bay of Bengal.

Sediments are tunnelled to the fan via the delta-front trough, the *Swatch of no Ground.* The Bengal Fan can be divided into three parts - upper fan, middle fan and lower fan. Most of the fluxes seem to get accumulated within the upper and the middle fan. The Bengal Fan is underlain by thick sequence of sediments derived from the peninsular and Himalayan rivers. More than one billion tons of sediments are discharged into the Bay annually, mostly from the G-B system (Milliman and Meade, 1983). Average sedimentation rate on the fan is 20-30 cm per 1000 years (Wetzel and Wijayananda, 1990).

In order to understand seasonal variations in source to sink pathways of fluvial influx into the Bay, from time series sediment trap observations, it had been inferred that concurrent with riverine flux maxima, the Bay witness a maximum terrigenous influx during the SW monsoon, the main source of which is the Himalayan rivers (Ittekkot et al., 1991). These fluxes dwindle during rest of the year. Therefore, it was thought that the influx of terrigenous sediments into the bay is broadly regulated by precipitation or melting pulses from the Himalayas (Ittekot et al., 1991). Moreover, a long distance dispersal of these sediments by low salinity Bay of Bengal waters into the southeastern Arabian Sea and along the equatorial region, mostly during the NE monsoon, has also been documented (Shetye et al., 1991; Chauhan and Gujar, 1996). The occurrence of high magnitude, short-lived, cyclones/depressions in the Bay is a common phenomenon during the NE monsoon (Chauhan, 1995), which brings torrential rains with sporadic fluvial discharge. Dispersal mechanism of the fluvial discharge into the Bay is, therefore, complex, especially during short (weekly) events associated with depression/cyclones or large spatial and temporal variability in the rainfall. In order to understand the dispersal pathways of the suspended sediments discharged by the G-B river system during the winter (NE) monsoon, a sequence of 64 TSM images (OCM) during October 1999-March 2001, synchronous sea truth data with the aid of salinity variations have been used to construct dispersal pathways of the surficial fluvial flux in the northern Bay. As part of in situ measurements about 100-150 litres of seawater were filtered for each station for acquiring about 200-300 mg of TSM during November of 2000 at eight stations along Orissa Coast (Fig. 15). Owing to very small amount of samples for clay mineral analysis a fraction of these samples were passed through a membrane filter of ~2 mm. Clays are identified and quantified using the methods of Biscaye (1965).

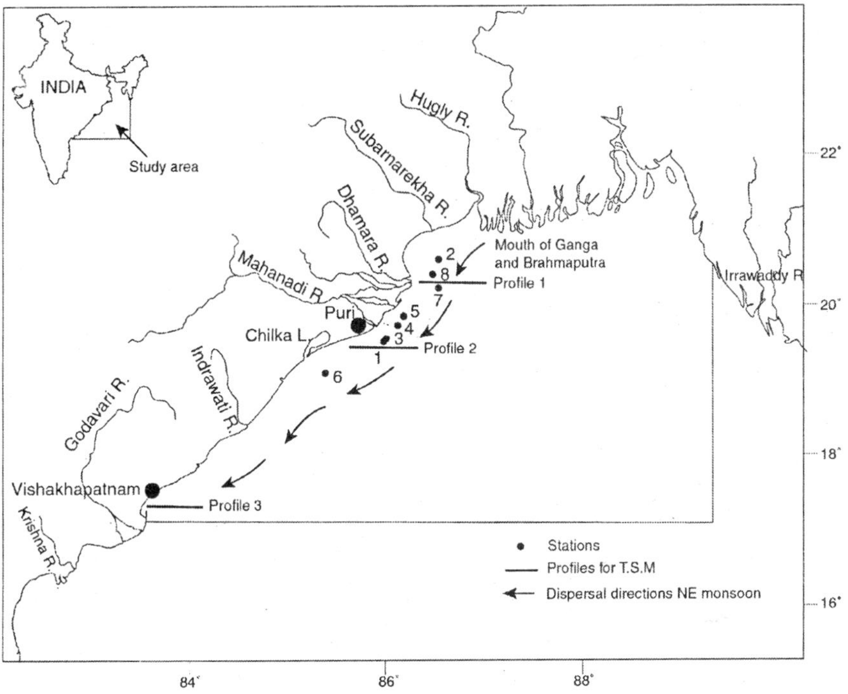

Figure 15: The study area and location map of the sampling stations for clay analysis. Sites of TSM profiles off Dhamara river, Chilika and Vishakhapatnam are also shown.

The generalised TSM patterns have large spatial and temporal variations during NE monsoon. During the month of October, there is gradual reduction in the spatial extents of the plumes of major rivers, except off the mouth of G-B system which has been traced for 120-160 km offshorewards. During the month of November, however, there are frequent, short duration pulses of high TSM. Gradually, the influx of TSM is reduced during December-January. The dispersal pattern derived from the sequential scenes and from the in situ TSM measurements is south-southwest.

The correlation between the sea truth and imageries data of TSM by and large is moderate ($r = 0.51$ $p = 0.001$) (Anuradha et al., 2000) because of complexities and inherent limitation of available algorithms. TSM values derived from the image are underestimated compared to in situ measurements in coastal waters. However, imagery derived dispersal patterns have high correlation with measured in situ currents magnitude and direction (Fig. 16). TSM patterns derived from the images in the study area (Fig. 15), though have limitation for the accurate quantification of fluxes, yet can be used in conjunction with measured TSM and current parameters for reconstruction of regional dispersal pattern of the fluvial flux, pending availability of an improved algorithm.

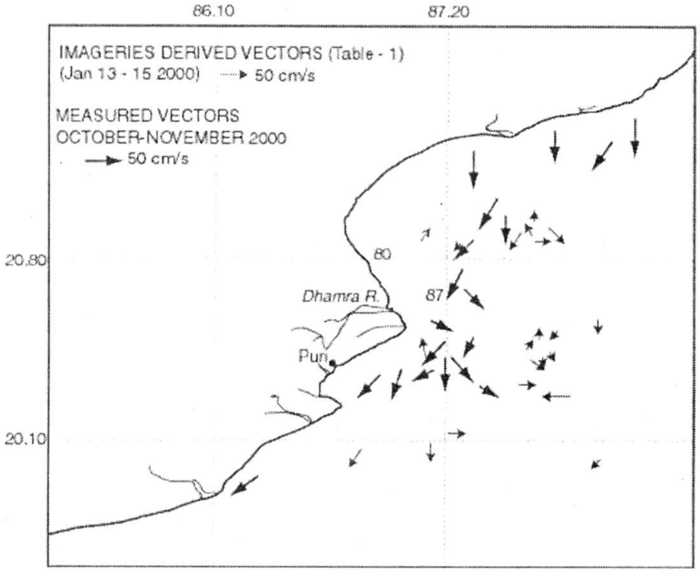

Figure 16: Surface currents estimated from sequential ocean colour images (in January 2000) and from ship board measurements (in October-November 2000) showing the equatorward advection along the coast.

The sequential variations in the TSM along the three W-E profiles along the northern, central and southern regions between 8 and 14 November, 1999 (Fig. 17) are also evaluated to further elaborate the pathways of TSM dispersal. On all individual profiles, from the north (off the Dhamara River) to the south (off Visakhapatnam), in the inland waters, there was a distinct localised area of high TSM, which decreases offshoreward on all the profiles, except on Profile 1 where a high TSM band existed 100 km offshore. On the subsequent profiles of 10 November 1999, in addition to the inland TSM enrichment, an additional band of high TSM, located about 50-100 km offshore was also observed. Between these two TSM enriched waters, lay the waters of reduced TSM values. These two waters, therefore, have different sources of TSM–deeper one from the pluses of the G-B system.

At all the profiles, the coastal waters have higher TSM. However, being in the vicinity of the G-B, the area off the Dhamara river continues to have much wider spatial extent and with higher TSM during the entire NE monsoon. In the offshore waters, however, the TSM is much reduced, except few isolated areas of high contents. It is, therefore, inferred that the fluvial discharge of the Himalayan rivers advects alongshore in narrow localised bands. The time lag of the advection of this TSM among profiles 1-3 was rather four days, which implies that dispersal rate is rather rapid (over 250 km in six days). These results distinctly suggest high dynamic nature of equatorward hydrography to disperse and distribute the fluvial influx along the shelf in short span of time.

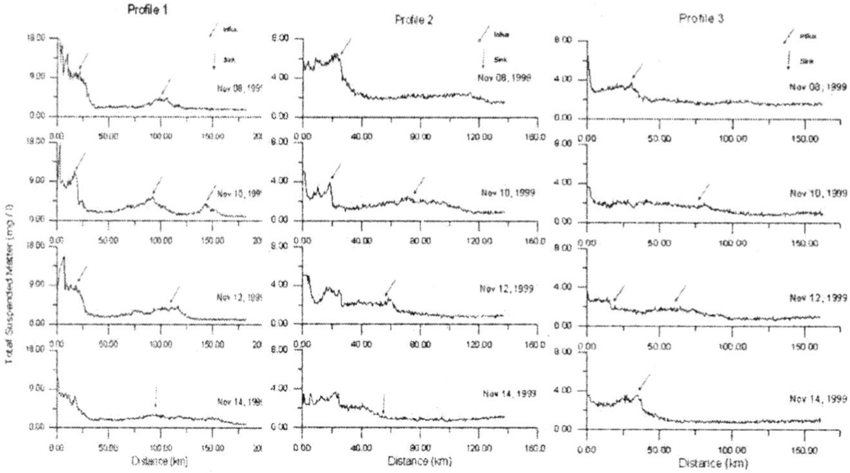

Figure 17: Weekly spatial and temporal variations in the TSM along W-E (Profile 1 off the mouth of Mahanadi-Dhamra shows the inland shift of plume in coastal waters during 8-14 November 1999. Profile 2 south of Paradeep shows sequential increase of TSM from offshore into inland waters and time lag of two days. Profile 3 at Visakhapatnam shows sequential delay in the enrichment of TSM in coastal waters). Arrows indicate influx and sink of the TSM.

Clay minerals transported in the seawaters were evaluated during November of 2000 which further elucidates the advection of G-B fluxes. Since the clay minerals (mean grain size <2 μm) (i) regulated by geology and drainage characteristics of catchment area, have fluvial source specific assemblages, and (ii) have a potential to be transported regionally in suspension, they have potential to be used as tracers of a specific source. The clays present in the surface waters along the study area are illite, chlorite, kaolinite, and smectite (traces) in the order of abundance (Table 3). Because the clay assemblage of illite, chlorite, and kaolinite are characteristics of the load of the G-B system (Konta, 1985), which is similar to one observed in the surface sea water

Table 3: Clay abundance in the suspended sediments (water depth 0-2 m) at selected locations (Refer Fig. 15 for station locations)

Station	*Location*	*Water depth*	*Illite*	*K+C*	*K/C*	*Smectite*
1	19° 40.1'N, 85° 50.1'E	22	61	17.9	1.58	5.1
2	21° 05.5'N, 87° 09.8'E	20.4	67	17.3	1.48	5.6
3	19° 40.1'N, 85° 50.2'E	32	61	16.9	1.51	5.4
4	20° 04.0'N, 86° 50.0'E	52.5	70.9	19.8	0.56	5.8
5	19° 55.1'N, 86° 28.4'E	32	73.1	15.5	0.61	6.4
6	19° 03.3'N, 85° 28.6'E	29.5	61.4	17.9	0.78	6.5
7	20° 42.6'N, 87° 19.1'E	28	69.4	19.8	0.63	5.1
8	20° 50.0'N, 87° 12.9'E	24.2	67.2	17.8	0.68	5.8

samples, it is inferred that fluvial flux of the G-B system is dispersed and distributed along the shelf by equatorward hydrography.

Considering the observed illite and chlorite in the surficial sediments along the continental shelf off Orissa, the study suggests that the source of chlorite is from the distributaries of the River Mahanadi. In general, chlorite (clay) is produced under arid, cold climate, and is mostly found in the load of Himalayan rivers G-B (Konta, 1985). The suspended load of the river Mahanadi and its distributaries is dominated by the smectite, kaolinite with minor illite and, therefore, the source of chlorite appears to be of not local origin. The integrated results of the present study, for the first time, distinctly suggest higher influence of the G-B system onto the coastal regions on the northern Bay of Bengal. The reduced terrigenous flux into the central and the northern traps of the Bay of Bengal during the NE monsoon, therefore, appears linked with the dispersal mechanism, rather than linearly related with magnitude of the flux from the G-B.

6. CONCLUSIONS

Surface oceanic circulation, besides turbidity currents and the bottom topography, affects the dispersal of materials within the Bay. Influx from the Himalayan and the Peninsular rivers reaches maximum by August; however, the perennial nature of the G-B system feeds the northern Bay even after the cessation of the summer monsoon. The seasonal trends of freshwater discharge in the Bay remain consistent over decades as evidenced from the climatology. However, the inter-annual variability of flux by the Himalayan rivers, unlike the Peninsular rivers, does not show one-to-one correlation with the Indian summer monsoon indices. The northern shelf of the Bay, being a macro-tidal region, remains ever-turbid due to the round the year discharge from the Himalayan rivers. The *Swatch of no Ground* is a prospective track to carry the sediment loads from the G-B source into the Bengal Fan and as far into the southern Indian Ocean; however, the dispersal of sediments along the Indian coasts needs further investigation. Satellite ocean colour data is extremely useful to trace the surface sediment dispersal pathways.

As a major contribution to understand and delineate the source-sink pathways of the G-B flux during off-summer monsoon periods, the combined experiment using in situ and remote sensing data confirms that: (1) during the NE monsoon the suspended sediment influx of the Ganga-Brahmaputra system influences the coastal processes with much higher spatial variability along the northern region, than hereto before thought and (2) the supply of the terrigenous sediments to the deeper offshore regions of the central Bay during the NE monsoon is not related with the magnitude of influx of the sediments into the Bay, but the prevalent dispersal by the hydrography. A schematic of the general dispersal of the spread of G-B flux is depicted in Fig. 18. During the NE monsoon, influx of the G-B moves N-S initially, off

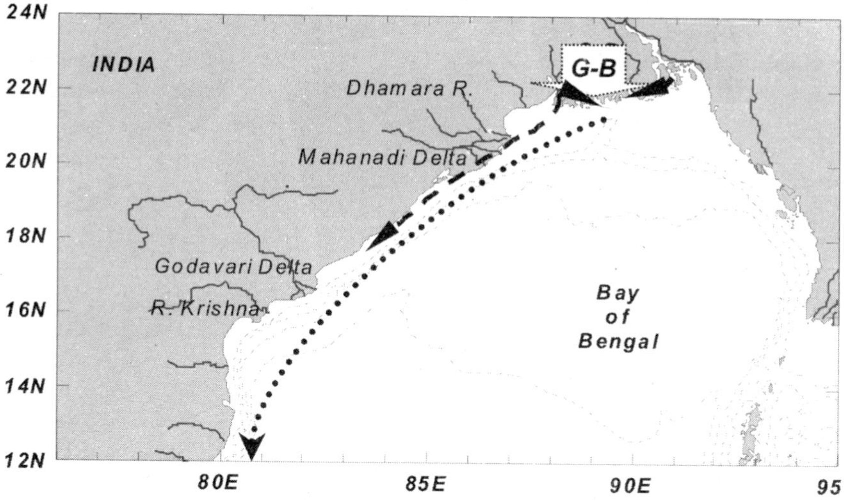

Figure 18: Conceptual depiction of sediment dispersal in the northern Bay. Solid and dashed arrows represent sediment pathways in the summer (SW) and the winter (NE) monsoon periods. The generalised surface current pattern along the east coast during the winter monsoon is shown with the dotted arrow. Bathymetry contours overlaid are for 200 m, 1 km, 2 km and 3 km from the coast.

the mouth, and thereafter advects southwest alongshore in the form of coastal sediment plumes, reducing the salinity of the coastal waters along the entire northern Bay during October-December. A strong relation exists between enhanced episodic discharges of the Ganga-Brahmaputra and augmented coastal turbidity during weekly events. It is observed that during short (weekly) events of very high pulse of TSM discharge by the G-B system, the fluvial fluxes do not advect offshoreward into the deeper offshore regions of the north-central Bay, but are transported alongshore and distributed along the shelf. These observations have implications for a possible different sink pathway, and biogenic processes associated with stronger/weaker monsoon activities.

REFERENCES

Anuradha, T., Suneethi, J., Dash, S.K., Pradhan, Y., Prasad, J.S., Rajawat, A.S., Nayak, S.R. and Chauhan, O.S., 2000. Sediment dispersal during NE monsoon over northern Bay of Bengal: Preliminary Results using IRS–P4 OCM data. *In: Proceedings of PORSEC*, **2:** 813-815.

Biscaye, P.E., 1965. Mineralogy and sedimentation of recent deep-sea clays in the Atlantic Ocean and adjacent seas and oceans. *Bull Geol Society of America*, **76:** 803-832.

Chauhan, O.S. and Gujar, A.R., 1996. Surficial clay mineral distribution on the south western continental margin of India: Evidence of input from the Bay of Bengal. *Cont. Shelf Res.*, **16(3):** 321-333.

Chauhan, O.S., 1995. Monsoon induced temporal changes in the beach morphology and associated sediment dynamics, central east coast of India. *J. Coastal Res.*, **11:** 776-787.

Cutler, A.N. and Swallow, J.C., 1984. Surface currents of the Indian Ocean (to 25°S, 100°E) compiled from historical data archived by the Meteorological Office, Bracknell, U.K., *Rep. 187*, Inst. of Oceanogr. Sci., Wormley, Surrey, England, 1-8.

Duing, W., 1970. The monsoon regime of the currents in the Indian Ocean, *IIOE Oceanogr. Monograph No. 1*, Univ. Hawaii, East-west Central Press, Honolulu, 68 pp.

Emmel, F.J. and Curray, J.R., 1984. The Bengal Submarine fan, North eastern Indian Ocean. *Geo-Marine Letters*, **3:** 119-124.

Ganapathi, P.N. and Murty, S.V., 1954. Salinity and temperature variations of the surface water off Visakhapatnam coast. *Andhra University Memoirs in Oceanography,* **1:** 125-142.

Garcia, C.A.E. and Robinson, I.S., 1989. Sea surface velocities in shallow seas extracted from sequential Coastal Zone Colour Scanner Satellite Data. *J. Geophys. Res.*, **94(C9):** 12681-12691.

Goodbred, S.L. and Kuehl, S.A., 1999. Holocene and modern sediment budgets for Ganges –Brahmaputra river: Evidence for highstand dispersal to floodplain, shelf, and deep-sea depocenters. *Geology*, **27(6):** 559-562.

Ittekkot, V., 1993. Particle flux studies in the Indian Ocean. *EOS, Trans. of the American Geophysical Union*, November 19, 1991, 527-530.

Ittekkot, V., Nair, R.R., Honjo, S., Ramaswamy, V., Bartsch, M., Manganini, S. and Desai, B.N., 1991. Enhanced particle fluxes in Bay of Bengal induced by injection of freshwater. *Nature,* **351:** 385-387.

Ittekkot, V., Safiullah, S., Mycke, B. and Seifert, R., 1985. Seasonal variability and geochemical significance of organic matter in the river Ganges, Bangladesh. *Nature*, **317:** 800-802.

Kolla, V., Henderson, L. and Byscaye, P.E., 1976. Clay mineralogy and sedimentation in the western Indian Ocean. *Deep-Sea Res.*, **23:** 949-961.

Kolla,V. and.Kidd, R.B., 1982. Sedimentation and sedimentary processes in the Indian Ocean. *In:* The Ocean Basins and Margins: The Indian Ocean, (eds. A.E.M. Nairn and F.G . Stehli), Plenum, New York, **6:** 1-45.

Konta, J., 1985. Mineralogy and chemical maturity of suspended matter in major rivers samples under SCOPE/UNEP, Transport of carbon and minerals in major world rivers. Part III (eds T. Degens and S. Kempe), Mitteilungen aus dem Geologisch - Paleontologischen Institut der Universitat Hamburg, pp. 569.

La Fond, E.C. and La Fond, K.G., 1968. Studies of Oceanic circulation in the Bay of Bengal. *Proc. Symp. Indian Ocean. Bull. Natl. Inst. Sci. India*, Vol I, 169-183.

LaViolette, P.E., 1967. Temperature, Salinity and Density of the World's Seas: Bay of Bengal and Arabian Sea. *Informal Rep. No. 67-57*, Naval Oceanographic Office, Washington D.C.

McCreary, J.P., Kundu, P.K. and Molinari, R.L., 1993. A numerical investigation of dynamics, thermodynamics and mixed-layer processes in the Indian Ocean. *Prog. Oceanogr.*, **31:** 181-244.

McCreary, J.P., Han, W., Shankar, D. and Shetye, S.R., 1996. Dynamics of East India Coastal Current, 2. Numerical solutions. *J. Geophys. Res.*, **101(C6):** 13,993-14,010.

Milliman, J.D. and M. Syvitski, J.P., 1992. Geomorphic/Tectonic control of sediment discharge to the ocean: The importance of small mountainous rivers. *J. Geol.*, **100:** 525-544.

Milliman, J.D. and Meade, R.H., 1983. World-wide delivery of river sediment to the oceans. *J. Geol.*, **91:** 1-29.

Murty, V.S.N., Sarma, Y.V.B., Suryanarayana, A.S., Babu, M.T., Santhanam, K., Rao, D.P. and Sastry, J.S., 1990. Some aspects of physical ocenography of the Bay of Bengal during the southwest monsoon. NIO Technical Report No. NIO/TR-8/90, National Institute of Oceanography, Goa.

Nath, B.N., Rao, V.P.C. and Becker, K.P., 1989. Geochemical evidence of terrigenous influence in deep sea sediments up to 8°S in the central Indian basin. *Mar. Geolo.*, **87:** 301-313.

Potemra, J.T., Luther, M.E. and Obrein, J.J., 1991. The seasonal circulation of the upper ocean in the Bay of Bengal. *J. Geophys. Res.*, **96(C7):** 12,667-12,683.

Ramaswamy, V., Kumar, B.V., Parthiban, G., Ittekkot, V. and Nair, R.R., 1997. Lithogenic fluxes in the Bay of Bengal measured by sediment traps. *Deep Sea Res. I*, **44(5):** 793-810.

Rao, D.P. and Murty, V.S.N., 1992. Circulation and geostrophic transport in the Bay of Bengal. *In:* Physical processes in the Indian Seas (eds G.N. Swamy, V.K. Das and M.K. Antony), Proc. First convention of the Indian Society for Physical Sciences of the Ocean, NIO, Goa.

Rao, D.P., 1977. A comparative study of some physical processes governing the potential productivity of the Bay of Bengal and Arabian Sea. *Ph.D. Thesis (Unpublished manuscript)*, Andhra University, Waltair, pp. 135.

Rao, D.P. and Sastry, J.S., 1981. Circulation and distribution of some hydrographic properties during the late winter in the Bay of Bengal. *Mahasagar, Bull. Natl. inst. Oceanogr.*, **14:** 1-15.

Ratna Reddy, S., Easton, A.K., Clarke, S.R., Narendra Nath, A. and Rao, M.V., 1995. Gyres off Somali coast and western boundary currents in the Bay of Bengal during the south-west monsoon. *Int. J. Remote Sens.*, **16(9):** 1679-1684.

Sanilkumar, K.V., Kuruvilla, T.V., Jogendranath, D. and Rao, R.R., 1997. Observations of the Western Boundary Current of the Bay of Bengal from a hydrographic survey during March 1993. *Deep Sea Res.*, Part I, **44(1):** 135-145.

Schott, F., Reppin, J., Fischer, J. and Quadfasel, D., 1994. Currents and transport of the Monsoon Current south of Sri Lanka. *J. Geophys. Res.*, **99:** 25,127-25,141.

Shankar, D., McCreary, J.P., Han, W. and Shetye, S.R., 1996. Dynamics of the East India Coastal Current 1. Analytic solutions forced by interior Ekman pumping and local alongshore winds. *J. Geophys. Res.*, **101(C6):** 13,975-13,991.

Shankar, D., Vinaychandran, P.N., Unnikrishnan, A.S. and Shetye, S.R., 2002. The monsoon current in the north Indian Ocean. *Progr. Oceanogr*, **52(1):** 63-119.

Shankar, D. and Shetye, S.R., 2001.Why is mean sea level along the Indian coast higher in the Bay of Bengal than in the Arabian Sea? *Geophysical Res. Lett.*, **28(4):** 563-566.

Shetye, S.R., Gouveia, A.D., Shenoi, S.S.C., Sundar, D., Michael, G.S. and Nampoothiri, G., 1993. The western boundary current of the seasonal gyre in the Bay of Bengal. *J. Geophys. Res.*, **98:** 945-954.

Shetye, S.R., Shenoi, S.S.C., Gouveia, A.D., Michael, G.S., Sundar, D. and Nampoothiri, G., 1991.Wind driven coastal upwelling along the western boundary of the Bay of Bengal during the southwest monsoon. *Continental Shelf Res.*, **11:** 1397-1408.

Subramanian, V., 1993. Sediment load of Indian rivers. *Curr. Sci.*, **64(11&12):** 928-930.

Tomczak, M. and Godfrey, J.F., 2001. Regional Oceanography: An Introduction, online version at *http://www.es.flinders.edu.au/~mattom/regoc/pdfversion.html.*

UNESCO, 1971. Discharge of selected rivers of the world, Vol. II and Vol. III, Paris.

Varadachari, V.V.R., Murty, C.S. and Das, P.K., 1968. On the level of least motion and the circulation in the upper layers of the Bay of Bengal. *Bull. Natl. Inst. Sci. India*, **38(I):** 301-307.

Varkey, M.J., Murty, V.S.N. and Suryanarayana, A., 1996. Physical Oceanography of the Bay of Bengal and Andaman Sea, *Oceanography and Marine Biology: an Annual Review* (eds A.D. Ansell, R.N. Gibson and Margaret Barnes), **34:** 1-70.

Vinaychandran, P.N., Masumoto, Y., Mikawa, T. and Yamagata, T., 1999. Intrusion of Southwest Monsoon Current into the Bay of Bengal. *J. Geophys. Res.*, **104:** 11077-11085.

Vorosmarty, C.J., Fekete, B.M. and Tucker, B.A., 1998. Global River Discharge Database (RivDIS) V. 1.1. Available online (http://www-eosdis.ornl.gov) and on CD-ROM from the ORNL Distributed Active Archive Center, Oak Ridge National Laboratory, Oak Ridge, TN, USA.

Wetzel, A. and. Wijyananda, N.P., 1990. Biogenicsedimentary structures in the outer Bengal Fan deposits drilled during leg 116. *Pro. of Ocean Drill Progress, Science Research*, **116:** 15-24.

Wyrtki, K., 1973. An equatorial jet in the Indian Ocean. *Science*, **181:** 262-264.

Wyrtki, K., 1971. Oceanographic atlas of International Indian Ocean Expedition. National Science Foundation, Washington, D.C., pp. 531.

Yu, L., O'Brien, J.J. and Yang, J., 1991. On the remote forcing of the circulation in the Bay of Bengal. *J. Geophys. Res.*, **96(C11):** 20,449-20,454.

Spatial Heterogeneity of Biogeochemical Parameters in a Subtropical Lake

Ilia Ostrovsky and Assaf Sukenik

Israel Oceanographic and Limnological Research
Kinneret Limnological Laboratory
P.O.Box 447, Migdal 14950, Israel
ostrovsky@ocean.org.il

1. INTRODUCTION

Spatial variability in aquatic ecosystems is caused by physical (upwelling, fronts, seiches, turbulence, inflows, etc.), chemical (e.g. unequal nutrient loads), biological processes (grazing, migration, etc.), and their interactions (Reynolds, 1984; Kalikhman et al., 1995; Knauer et al., 2000). Boundary processes can generate intense resuspension, turbulent mixing, and upward fluxes of nutrients (Imboden and Wuest, 1995; Ostrovsky et al., 1996; Macintyre et al., 1999). Presently the role of various phenomena in formation of spatial variability of suspended particles and chemical fluxes is poorly understood. The "visualization" of the complex spatial organization of an aquatic ecosystem is an essential step toward the identification of the processes responsible for its formation. Analysis of such information can help to reveal mechanisms that control eutrophication and determine local importance of internal and external loading of nutrients. Revelation of seasonal aspects of spatial heterogeneity can be also useful for verification of the existing complex three-dimensional mathematical models (e.g. Hodges et al., 2000) of aquatic ecosystems.

In most inland water bodies the patchy distribution of any parameter or process cannot be recorded since data is collected from only a few sampling stations. Intensification of the traditional monitoring scheme by adding stations and increasing sampling frequency could resolve this problem but would require substantial capital investment in manpower, equipment and operational costs. On the other hand, ignoring the patchy structure and processes would restrain our knowledge about the organization and functioning of the ecosystem.

An innovative approach to increase spatial resolution of several in situ measured limnological parameters (temperature, turbidity, conductivity, and chlorophyll) was recently implemented in Lake Kinneret (Israel) by operating an underwater towed undulating monitoring system (U-TUMS, Sukenik et al., 2002). The purpose of this operation was to study, with a moderate effort and within a limited time schedule, spatial distribution of limnological parameters and to identify processes that occur over a large portion of the water body. Herein we describe some results of the U-TUMS near to synoptic monitoring of Lake Kinneret to identify the spatial heterogeneity of water quality parameters during different seasons in relation with limnological processes that may be responsible for such patterns.

2. LAKE KINNERET (STUDY SITE)

Lake Kinneret (Sea of Galilee) is a warm-monomictic lake located at about 210 m below mean sea level in the northern part of Israel. The lake is 22 km long and 12 km at maximum width and its surface area is 168 km^2. Its maximum and mean depths are ~40 m and 22 m, respectively. The lake is stratified during most of summer and autumn, with surface temperatures approaching 30°C with anoxic and cooler hypolimnion. The main water inflow (60-70% of the total) is the River Jordan, while the main outflow is by pumping through Israel's National Water Carrier, supplying nearly 50% of all drinking water in the country. Lake Kinneret serves also as a vacation and recreation centre and used for commercial fish harvest. Thus, the conservation of Lake Kinneret ecosystem for future generations is a subject of national mission.

In Lake Kinneret, physical, chemical and biological processes are highly dynamic and spatially heterogeneous, thus make monitoring of the entire system very complex (Nishri et al., 2000; Sukenik et al., 2002). The heterogeneity of ecosystems arises from various processes: thermal fluxes, boundary processes, stratification, water motions, schooling of biological populations, etc. In Lake Kinneret both surface and internal waves cause resuspension of particles and their uneven distribution in littoral and sublittoral zones (Ostrovsky et al., 1996; Ostrovsky and Yacobi, 1999). Daily wind-produced internal seiches are common from May through September. Interaction of the seiches with bottom induces diapycnal mixing the sublittoral zone. This creates favourable conditions for enhanced local productivity of planktonic community (Ostrovsky et al., 1996). In addition, large loads of nutrients and suspended particles enter the north zone of the lake during the winter with floods, unevenly distributed by currents (Ostrovsky and Yacobi, unpublished) and support the patchiness development of the plankton community.

3. MATERIAL AND METHOD

Several water quality parameters were collected using the underwater towed undulating monitoring system (U-TUMS), This system is comprised of a vehicle (carrier), a set of sensors and navigation devices (geographic positioning system—GPS, SONAR and speedometer) which are interconnected and operated via an on-board computer. The U-TUMS is towed behind the boat along chosen trails (Fig. 1). The computer operates the vehicle's steering mechanism and causes the vehicle to undulate from near-surface to near-bottom or to the depth of 27 m (if the bottom positioned at deeper depth). Data from sensors were collected on a computer and presented on screen in real time. The underwater vehicle (MiniBAT, GuildLine, Canada) was loaded with: (1) a multi-sensor probe CTD that measures water electric conductivity, temperature and depth (Applied Microsystems Limited, Canada); (2) an optical backscatterance sensor that measures water turbidity (OBS-3, D&A Instrument Company, USA) and (3) a fluorometer for sensing chlorophyll (MinitracaII, Chelsea Instrument, UK). The fast time response of these sensors assured accurate measurements at a distance resolution of less than 15 cm within the water column, while the vehicle was towed at a speed of 4.5 knots and undulating (diving or climbing) at rates of upto

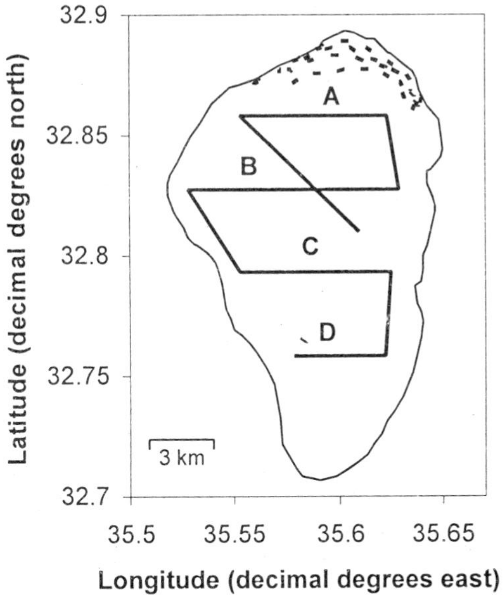

Figure 1: Lake Kinneret contour and trails of surveys. Solid bold line is a standard U-TUMS trail designed to monitor water quality parameters; A, B, C, and D are north-west sections of this trail. Dotted line is a trail of a survey aimed to study the dispersion of Jordan River water in the Northern part of the lake on 27 June 2000.

90 cm s^{-1}. The collected data is then analyzed to produce a two-dimensional view (vertical by horizontal) of the U-TUMS transect. Subsequently, several such two-dimensional transects could be combined to produce a three-dimensional distribution pattern for a given parameter. The horizontal (areal) variability of each parameter is then extracted and presented for a given water depth layer to reflect spatial variations.

The collected data was filtered for errors (based on logic limits and expected values) and transformed from electronic values to physical quantities. Electric conductance is presented in practical salinity units (PSU) assuming a relatively stable ion composition (Millero, 1996), OBS turbidity measurements (in relative turbidity units, RTU) were transformed to NTU (nephelometric turbidity units) based on linear transformation (NTU = 0.102 RTU + 6.56) defined by a calibration procedure, and the fluorescence measurements were transformed to chlorophyll concentrations according to the linear relationship based on direct chlorophyll measurements of water samples analyzed by the U-TUMS fluorometer. The data were integrated into a pseudo-three-dimensional presentation using Surfer 7.0 (Golden Software Inc.) applying near-neighbour interpolation procedure.

4. RESULTS AND DISCUSSION

4.1 Spatial Heterogeneity of Stratified Lake

A typical large-scale survey carried out in spring, when Lake Kinneret became stratified, demonstrated spatial variability of the thermal structure (Fig. 2). The upper warmer water layer, the epilimnion, is distinctly separated from the deeper cool water layer, the hypolimnion, by a relatively thin water layer, the metalimnion layer (see position of 16.5°C to 19°C-isotherms). During the sampling time, which takes up to one hour for a single west to east transect, one can detect the variation of the vertical distance between various metalimnetic isotherms with the location. The vertically stretched and constricted sections of the metalimnetic isotherms show the presence of the second basin-scale vertical modes of the internal waves. The tilting or bowing of the entire metalimnion is indicative for the presence of the first modes of the basin-scale internal seiches (Imberger and Patterson, 1990; Imboden and Wuest, 1995). The rapid performance of the U-TUMS allowed the identification of the thermal signature of the dominant basin-scale internal seiches, which in Lake Kinneret have cycles of 12 and 24 hours (Antenucci et al., 2000), which are much longer than the sampling time of a single transect.

In July and August surveys we detected a distinct difference in surface temperature in various parts of the lake. The highest surface water temperature was found near the eastern part of the lake. The difference in water surface temperatures was concurrent with higher concentrations of turbidity and chlorophyll and was apparently related to higher heat absorption of the

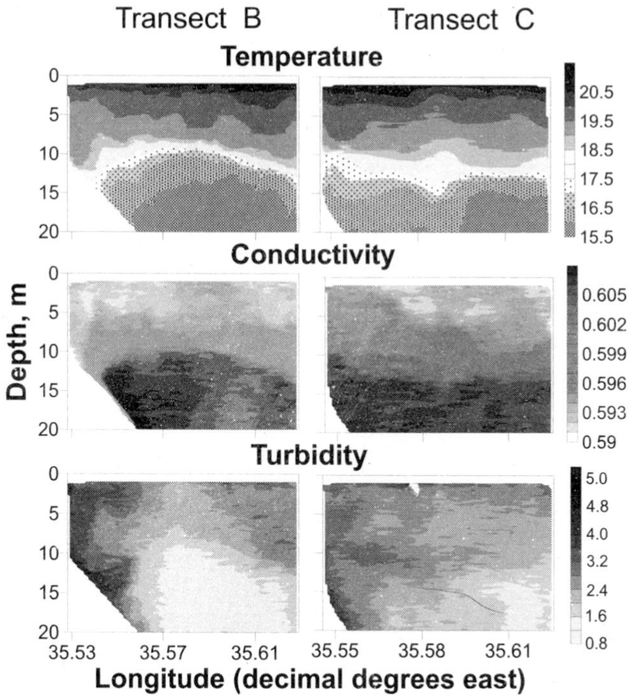

Figure 2: Vertical distributions of water temperature (°C), conductivity (PSU), and turbidity (NTU) along two transects at the beginning of stratification (27 March 2001). Positions of transects C and D are shown in Fig.1

highly turbid waters. In addition, frequently observed algal patches in Lake Kinneret (e.g. Yacobi et al., 1993; Gitelson et al., 1994) can be caused by western wind, as positively buoyant phytoplankton species can be accumulated near the downwind (eastern) shore or negatively buoyant algal species can be carried by subsurface returned currents and concentrated near the western shores. These patches can then drift and disperse throughout the lake via wind-induced convection cells and large gyres. At the earlier stage of stratification (Spring 2001, Fig. 2), some conductivity isolines crossed the isotherms in shallower regions suggesting diapycnal mixing. In lakes, most of the mixing of the hypolimnetic nutrient-rich water with epilimnetic water occurs at slopped bottom in sublittoral zone due to breaking of internal waves and enhanced turbulence (Imboden and Wuest, 1995; Sakai et al., 2002). Such mixing implies also an enhanced upward flux of nutrients via the thermocline which, in turn, should create favourable conditions for local increase in productivity (Ostrovsky et al., 1996, Macintyre et al., 1999). Since direct measurements of limiting nutrient concentrations in the upper productive layer are often problematic due to their rapid uptake by algal and bacterial cells, analysis of spatial distribution of phytoplankton standing stock (e.g. by means of chlorophyll concentration) can pinpoint the locations where nutrient enrichment is most profound at various seasons. In stratified mesotrophic Lake Kinneret, where primary productivity is strongly controlled by nutrient availability (bottom-up regulation), the location of the small but persistent injections of nutrients via metalimnion (diapycnal transport) can be discovered by increased chlorophyll concentrations. In particular, increased chlorophyll concentrations were detected in the marginal zones, where diapycnal mixing is most active (Ostrovsky et al., 1996).

Along the cross-lake transects high turbidity was usually observed at the peripheral areas, whereas the central part of the lake was less turbid (e.g. Fig. 2). The higher peripheral turbidity is caused by resuspension of particles by surface and internal waves (Ostrovsky and Yacobi, 1999; Yacobi and Ostrovsky, 2000). The intrusion of turbid layer within the metalimnion (see turbid tongue at ~15 m depth near the western shore, Fig. 2, transect B), supports our hypothesis that resuspended particles can be transported from the shallower areas toward the lake centre by rapid metalimnetic jets (Ostrovsky and Yacobi, 1999). Below the thermocline a layer of turbid water overlies the sloped bottom (Fig. 2, transects B and C). This turbid benthic boundary layer is produced by resuspension of bottom sediments due to strong shear stress induced by near-bottom seiches-related currents and by breaking of steep internal waves at the sloped bottom (Lemckert et al., 2004). High turbidity is also typical for the near-shore epilimnetic zone, where littoral benthic particles are resuspended due to energetic surface waves and/or shear stress inducing by along-shore currents.

4.2 SPATIAL HETEROGENEITY OF NON-STRATIFIED LAKE

In the winter as the lake destratified, temperatures across the water column at particular geographical location are normally nearly identical (Fig. 3), indicating that the lake is well-mixed vertically. During sunny winter days the upper 3-4 m layer can be warmed up and its temperature can be slightly higher than that of the underlying layer. Despite vertical homogeneity of the water quality parameters, the lateral heterogeneity can be explicitly observed (lateral variations in temperature, conductivity, and chlorophyll over the western, central, and eastern parts of the lake can be seen on Figs 3 and 4). The lateral heterogeneity in winter can be the result of temporal separation of large parts of the lake. Such large-scale water motions (e.g. gyres, currents) may isolate to some extent different lake areas from each other.

As the lake destratified, relatively narrow patches of high turbidity were vertically and horizontally distributed chaotically over the entire water volume (Fig. 3). This pattern was typical to the unstable conditions of the winter characterized by low air temperatures, strong winter storms with high wind velocity, rains and river floods. It suggests the possible involvement of

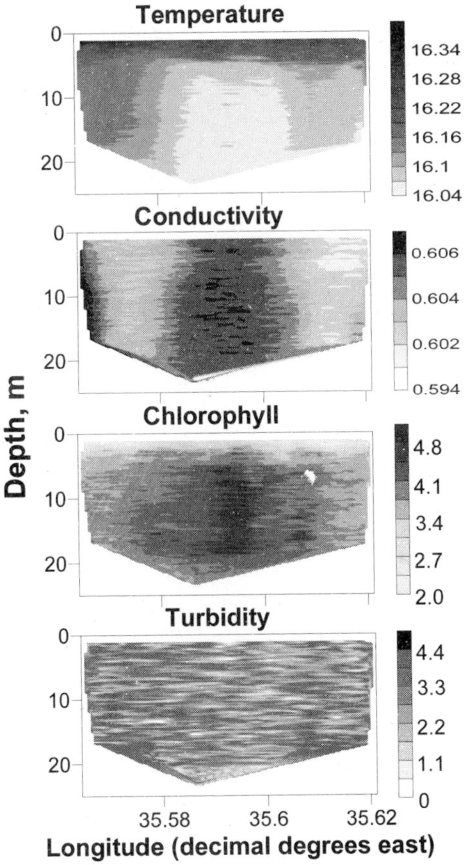

Figure 3: Vertical distributions of water temperature (°C), conductivity (PSU), chlorophyll concentration (µg l^{-1}), and turbidity NTU) along transect D on 18 February 2001 (during the complete holomixis). Position of transect D is shown in Fig. 1.

processes that lead to local agglomeration or concentration of particles (planktonic or inorganic substances).

In mid-February the mean section of the cross-lake transects had the highest chlorophyll concentrations. The increased algal concentration in the central part of the lake was apparently detected after the abrupt mixing of the residual (near-bottom) phosphorous-rich hypolimnetic water with upper productive strata just prior to the holomixis. Such an addition of the major nutrient component could stimulate local algal productivity. Another source of phosphorous could be bottom sediments that had been accumulated at the deepest central part of the lake under anoxic conditions prior to the holomixis. Since the lake interior was apparently separated by a basin-scale gyre from the peripheral areas, the higher algal biomass was observed in the lake centre during some weeks after the establishment of the overall holomixis (Fig. 4). Higher turbidity near the beaches, as a result of resuspension of benthic particles in the shallow parts of the lake, also was the pronounced feature during the winter time. Temporal isolation of the shallow zones confined dispersion of the turbid water.

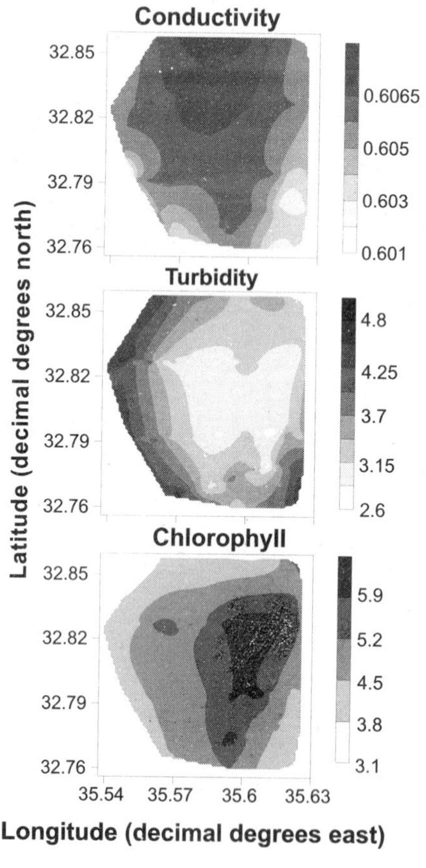

Figure 4: Spatial distributions of conductivity (PSU), turbidity (NTU), and chlorophyll concentration ($\mu g\ l^{-1}$) at the northern and central parts of Lake Kinneret on 18 February 2001. Data were averaged over upper 10 m water stratum.

4.3 DISPERSION OF JORDAN RIVER IN THE UPPER LAKE STRATUM

In order to investigate the dispersion the Jordan River water in the upper productive stratum and the impact of the incoming water on limnological parameters, a survey was conducted at a confined area at the northern part of Lake Kinneret in June 2000 (survey track is shown in Fig. 1). During that time the river flow was about four times lower than that in winter (the rainy season). Dispersion of the riverine water in the lake body can be followed by salinity changes, since the salinity of the lake water is an order of magnitude higher than that of the Jordan River. The lowest salinity and the highest turbidity were found at the top northern part of the lake, just near the entrance of the Jordan River (Fig. 5). A small patch of the cooler water can be recognized at the topmost location (the river mouth), indicating that temperature of the entering water was slightly lower than that of the water of the upper stratum, which is often the case. The signals of the lower salinity (conductivity) and enlarged turbidity were diverged toward the western part of the lake, which indicates the anti-clockwise circulation of water. This kind of water movement has been detected in summer (Serruya, 1975). Higher turbidity promotes a greater absorption of light energy by the surface water and its notable heating. This perfectly explains the performance of a tongue of warmer water that follows the dispersion of turbid Jordan water at the lake surface. Distribution of the nutrient-rich water that enters the lake from the Jordan River also excites the algal development at the northern area of Lake Kinneret (Ostrovsky and Yacobi, unpublished results).

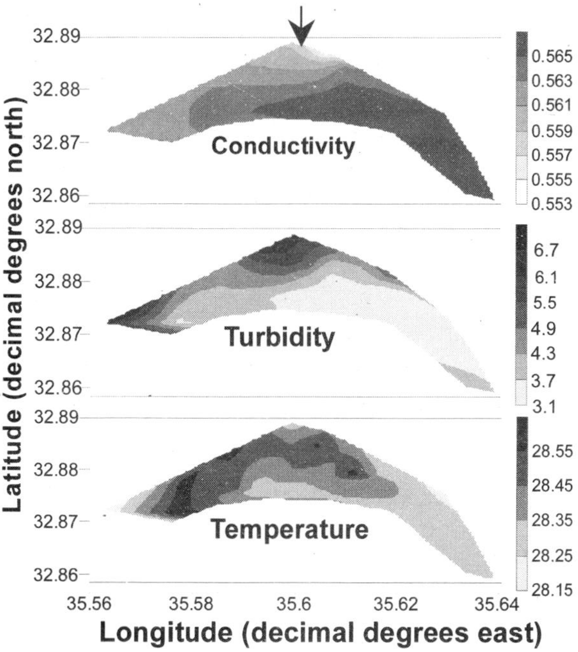

Figure 5: Spatial distributions of conductivity (PSU), turbidity (NTU), and temperature (°C) near the Jordan Rivet outlet area on 27 June 2000 (period of low water inflow). Data were averaged over the upper 5 m water stratum. Survey track is shown in Fig. 1. The Jordan River outlet area is shown by arrow.

At the end of the flood season 2001, the dispersion of Jordan water in the upper stratum of the lake can be clearly seen by dispersion of water of lower conductivity from the data obtained during a standard survey (Fig. 6). The distribution of the fresher water at the end of March 2001 suggested that clockwise circulation was predominant. Serruya (1975) observed this type of water circulation during the period of weak wind (< 3 m sec^{-1}) in winter and spring. As seen in Fig. 6, high chlorophyll concentrations corresponded to water of lower conductivity, which is a signal of water from the Jordan River, which is the main supplier of allochthonous nutrients (phosphorous and nitrogen). In other words, the addition of the nutrient-rich Jordan water to the upper productive strata of the lake could stimulate local development of algae. This observation also agrees with the direct measurements of algal productivity in areas affected by intrusion of the Jordan River (Yacobi, personal communication).

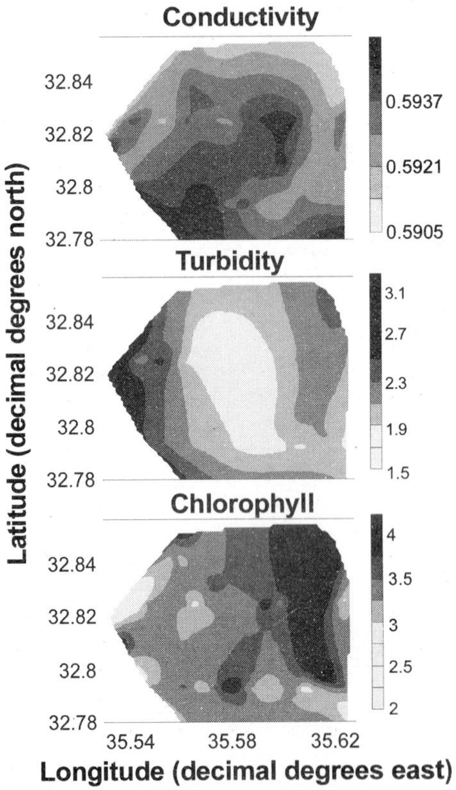

Figure 6: Spatial distributions of conductivity (PSU), turbidity (NTU) and chlorophyll (µg l^{-1}) near the central and northern parts of Lake Kinneret on 27 March 2001 (at the end of flood period). Data were averaged over the upper 5 m water stratum.

5. CONCLUSIONS

The monitoring of inland waters must take into consideration the spatial heterogeneity of many parameters. Intensification of the current monitoring programme in Lake Kinneret by regular operation of U-TUMS significantly improved the capabilities to identify spatial distribution of limnological parameters. The spatial heterogeneity in distribution of the main water quality parameters allowed revealing the underlying physical and biological processes. In particular, the distribution images of temperature, conductivity, chlorophyll, and turbidity indicated the importance of resuspension process from the lake floor, "injection" of nutrient-rich water, and water movement in Lake Kinneret. Further understanding of the factors affecting spatial distribution referred from the collected data could lead to improved awareness of the relationships between external lake perturbations (i.e., lake management) and ecosystem attributes (e.g., water quality).

Acknowledgements

This research was financially supported by Israel Water Commission and partially by grants from the German–Israel Foundation (contract no. I-711-83.8/2001). The authors thank Ms. M. Shlichter for initial data analysis and the crew members of RV Hermona: J. Eston, M. Hatab, Z. Rozenberg, and N. Koren for technical assistance.

REFERENCES

Antenucci, J.P., Imberger, J. and Saggio, A., 2000. Seasonal evolution of the basin-scale internal wave field in a large stratified lake. *Limnol. Oceanogr.*, **45:** 1621-1638.

Gitelson, A., Mayo, M., Yacobi, Y.Z., Parparov, A. and Berman, T., 1994. The use of high-spectral-resolution radiometer data for detection of low chlorophyll concentrations in Lake Kinneret. *J. Plankton Res.*, **16:** 993-1002.

Hodges, B.R., Imberger, J., Saggio, A. and Winters, K.B., 2000. Modeling basin-scale internal waves in a stratified lake. *Limnol. Oceanogr.*, **45:** 1603-1620.

Imberger, J. and Patterson, J.C., 1990. Physical limnology. *Adv. Appl. Mech.*, **27:** 303-475.

Imboden, D. and Wuest, A., 1995. Mixing mechanisms in lakes, p. 83-138. *In:* A. Lerman, D. M. Imboden and J. R Gat [eds.], Physics and chemistry of lakes. Springer-Verlag.

Kalikhman, Y., Ostrovsky, I., Walline, P., Gophen, M. and Yacobi, Y.Z., 1995. Distribution fields for aquatic ecosystem components: method of identification of correlation zones. *Freshwater Biol.*, **34:** 317-328.

Knauer, K., Nepf, H.M. and Hermond, H.F., 2000. The production of chemical heterogeneity in Upper Mystic Lake. *Limnol. Oceanogr.*, **45:** 1647-1654.

Lemckert, C., Antenucci, J., Saggio, A. and Imberger, J., 2004. Physical properties of turbulent benthic boundary layers generated by internal waves. J. *Hydraulic Engineering*, **131**: 58-69.

Macintyre, S., Flynn, K.M., Jellison, R.J. and Romero, J.R., 1999. Boundary mixing and nutrient fluxes in Mono Lake, Califoirnia. *Limnol. Oceanogr.*, **44**: 512-529.

Millero, F., 1996. Chemical Oceanography, CRC Press, Boca Raton, Fl, 469 pp.

Nishri, A., Imberger, J., Eckert, W., Ostrovsky, I. and Geifman, J., 2000. The physical regime and the respective biogeochemical processes in the lower water mass of Lake Kinneret. *Limnol. Oceanogr.*, **45**: 971-981.

Ostrovsky, I., Yacobi, Y.Z., Walline, P. and Kalikhman, Y., 1996. Seiche-induced water mixing: Its impact on lake productivity. *Limnol. Oceanogr.*, **41**: 323-332.

Ostrovsky, I. and Yacobi, Y., 1999. Organic matter and pigments in surface sediments: Possible mechanisms of their horizontal distributions in a stratified lake. *Can. J. Fish. Aquat. Sci.*, **56**: 1001-1010.

Reynolds, C.S., 1984. The ecology of freshwater phytoplankton. Cambridge University Press, 384 pp.

Sakai, Y., Murase, J., Sugimoto, A., Okubo, K. and Nakayama, E., 2002. Resuspension of bottom sediment by an internal wave in Lake Biwa. Lakes & Reservoirs. *Research and Management*, **7**: 339–344.

Serruya, S., 1975. Wind, water temperature and motions in Lake Kinneret: General pattern. *Verh. Internat. Verein. Limnol.*, **19**: 73-87.

Sukenik, A., Ostrovsky, I. and Nishri, A., 2002. Advanced approach for synoptic monitoring of a lake ecosystem: Lake Kinneret as a model. *In:* Preserving the quality of our water resources (H. Rubin, P. Nachtnebel, J. Fuerst, U. Shamir, eds.), Springer-Verlag Publishers, pp. 165-176.

Yacobi, Y.Z., Kalikhman, I., Gophen, M. and Walline, P., 1993. The spatial distribution of temperature, oxygen, plankton and fish determined simultaneously in Lake Kinneret, Israel. *J. Plankton Res.*, **15**: 589-601.

Yacobi, Y.Z. and Ostrovsky, I., 2000. Lake Kinneret sediments: spatial distribution of chloropigments during holomixes. *Arhiv fur Hydrobiologie*, **55**: 457-469.

Mixing and Internal Waves in a Small Stratified Indian Lake: Subhas Sarobar

N.R. Samal, Klaus D. Jöhnk[1], Frank Peeters, Erich Bäuerle[2] and Asis Mazumdar[3]

Limnologisches Institüt, University of Konstanz, Germany
Nihar.Samal@uni-konstanz.de, niharwre@sify.com
[1]Aquatic Microbiology/Institute for Biodiversity and Ecosystem Dynamics University of Amsterdam, The Netherlands
[2]Wasseransichten, Germany
[3]School of Water Resources Engineering, Jadavpur University, Kolkata, India

1. INTRODUCTION

The formation and break-up or erosion of stratification is a major process in all natural and man-made lakes, which controls to a large extent the functioning of their ecosystems. The key challenge here is the specific role of shear generated and convective turbulence in the formation/destruction of stratification and their interactions with internal waves. Three aspects of hydrodynamics, i.e. the dynamical state, the physical mechanism and the energy level are very important elements in understanding problems of turbulent mixing and the formation of vertical thermal stratification. The result of the stratification is the formation of the seasonal thermocline.

Mixing dynamics is of great importance for the management of water quality because it allows for water pollution prediction in relation to thermal stratification. In particular vertical mixing is an important aspect with respect to the exchange processes of heat and dissolved substances between the different vertical layers of the lake. It is controlled by atmospheric forcing at the water surface and the resultant advective and oscillatory motions in the water column. Generally speaking, the intensity of vertical mixing decreases with depth. Traditionally, mixing is expressed in terms of a turbulent mixing coefficient in analogy to the molecular diffusion coefficient. The turbulent coefficients are a measure of the intensity of the process and may subsequently be used in calculations related to water quality and heat transport. Mixing can be assessed either directly from microstructure measurements of the turbulence levels or indirectly by measuring temporal changes of the spatial gradients of some particular properties such as temperature.

In the present investigation, the seasonal cycle of thermal stratification has been simulated for a period of more than five years based on hourly meteorological input with a one-dimensional (vertical) turbulence model (LAKEoneD). Using these temperature-depth (TD) profiles, the annual cycle of phase velocities of long internal waves is calculated depending on the density difference in the stratified lake. Based on the knowledge of the actual value of the phase velocity the spectra of free internal oscillation can be calculated. That is done for the periods of significant stratification (April to November) by using a horizontally two-dimensional eigen-frequency model. From the calculations it becomes clear that the first order baroclinic free oscillations have periods significantly smaller than diurnal or semi-diurnal values. Thus, resonance with meteorological forcing on that time scale can be ruled out. Changing spatial patterns of observed temperature distributions indicate that the small shallow lake in the tropics exhibits differential mixing which may cause horizontal exchange flows between different stations.

2. LAKE CHARACTERISTICS AND DATA

Study site: Subhas Sarobar (Latitude 22° 34′-22° 34′ 30″ N and Longitude 88° 24′-88° 24′ 30″ E), under the administrative control of Kolkata Improvement Trust (KIT), represents the lung of East Kolkata with massive environmental fillip. The lake ecosystem is playing a key role in maintaining the oxygen balance and is also being used for sports, recreational and cultural activities. The vast water body and its two islands, one small and the other big, constitute an excellent habitat for diverse species of life and also have got potential for attracting the tourists. The water area of Subhas Sarobar (Fig. 1) is about 16.0 ha. The maximum length and width of the lake are

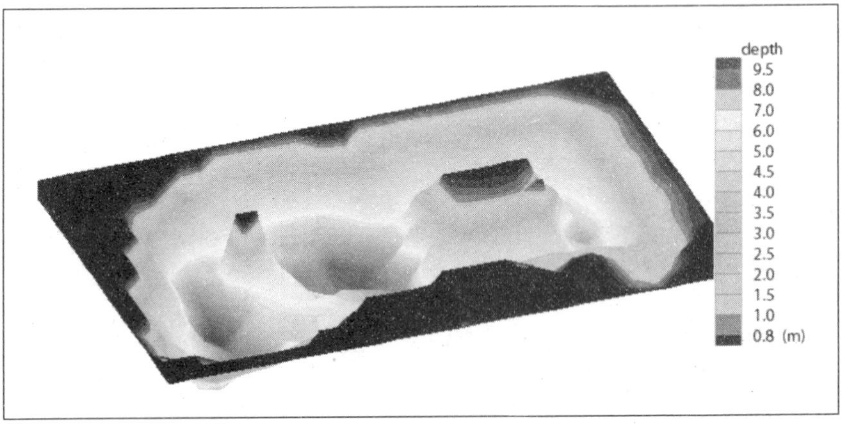

Figure 1: Three-dimensional view of the topography of lake, Subhas Sarobar, Kolkata, India.

617 m and 352 m respectively while the maximum depth and the mean depth of the lake are 10 m and 4.8 m, respectively. The waterbody is given on rent to the Department of Fisheries of Government of West Bengal state. The open space in the lake is an oasis in the space-limited city. Recently, the pressure of human activities on the Subhas Sarobar has increased manifolds. Over three thousand of people per day are using it for washing of clothes and utensils and for bathing.

Lake temperature data: The observation sites are selected considering the maximum water depth along the mid-reach (Station I, II and III) which are also least influenced by human intervention, to study the thermal stratification. The measurement of temperature profiles in this lake have been started during the month of May, 2003 using a WTW temperature/oxygen meter (Germany) along the depth at an interval of 0.5 m below the water surface, thus providing detailed information on the temperatures prevailing in all depth regions of the lake. All temperatures are considered to be accurate to within ± 0.1°C. Also some temperature profiles during the year 1999 are recorded by the Institute of Wetland Management and Ecological Design, Kolkata (Saha, 2000).

Meteorological data: The daily meteorological data, i.e. maximum and minimum air temperature, maximum and minimum relative humidity, mean wind speed and mean cloud cover were supplied by the Dumdum meteorological station (Kolkata), located within 9 km of the northern lake shore and also from the internet site Weather Underground (Pawson, 2004). In order to take adequate account of the diurnal cycle in meteorological forcing, which may be important for the long-term development of the thermal structure of the lake, the meteorological data are interpolated at hourly intervals. It is demonstrated that the numerical lake model is able to simulate the main features of the thermal structure of a lake without the luxury of accurate, high-resolution input data. The empirical equations for the interpolation of the meteorological data into hourly values are described somewhere else (Samal, 2004). The hourly values of the global irradiance are calculated depending on the course of the sun from the geographical position, cloud cover, C, and wind speed, U.

Light extinction coefficient: As the solar radiation is absorbed within the water column rather than at the air-water interface, the vertical distribution of heat depends not only on vertical mixing processes, but also on light extinction coefficient. The extinction coefficient for clear water is determined by the empirical formula given in Joehnk and Umlauf (2001)

$$K_{cw} = 1.7/Z_{secchi}$$

From the maximum measured Secchi-depth of Z_{secchi} = 1.7 m during our observation period, we infer a clearwater extinction of K_{cw} = 1.00 m^{-1}.

3. SIMULATION MODELS

3.1 Description of the Turbulence Model

The two-layer model: Turbulent kinetic energy in a lake is generated by the wind shear stress that acts at the lake surface due to meteorological activity above the lake. This causes a vertical mixing within the water column and therefore influences not only the temperature distribution but also the average residence time of the planktonic organisms in the euphotic zone and thus determines their survival conditions. As compared to the vertical variation of the dominant processes the horizontal water motion can be ignored if long-term processes and global dynamics are examined. In this case it suffices to establish a model for an arbitary offshore water column of specified depth, valid for an arbitrary lake.

Every temperature and density profile corresponds to a particular value of potential energy of the water column, which follows from the centre of gravity of this water column. The considered physical processes in a lake essentially amount to a mutual exchange between this potential energy and the turbulent kinetic energy (TKE) caused by wind shear stress and surface cooling. The centre of gravity of a stable density profile always lies below the centre of gravity of a homogeneous profile. Thus, kinetic energy must be expended to homogenize the stratified column or part of it, and conversely, an unstable water column neutralises itself automatically by transforming potential energy into turbulent kinetic energy (which in turn may homogenize a sublaying stable profile so as to re-transform itself via convection again into potential energy).

Heat flow over the surface of the domain and volume of production of heat by solar irradiance change, in general, both the location of the centre of gravity of a water column as well as its heat content. The mechanical wind energy induced into the lake by turbulence activity and the mixing of the water by convection, however, only change the potential energy, maintaining the heat content of the water column.

The vertical transport of the TKE that is generated by the surface wind shear or the cooling of surface water is substantially simplified by the two-layer assumption: according to this simplification the entire TKE is at all times uniformly distributed in the mixed layer, whilst the underlying water remains at rest. By this drastic assumption the TKE profile, that decreases rather exponentially, is averaged in the upper and neglected in the lower layer, so that the model becomes 'zero-dimensional' in respect to the TKE distribution. This two layered model was theoretically proposed and experimentally verified by Kraus and Turner (1967) and was further developed, e.g., by Niiler (1977) to simulate the thermal stratification in the ocean. The actual interface between the two layers, the diurnal turbocline, may be identified with the largest vertical gradient of the TKE in real lakes.

Its location does not, in general, agree with the location of the thermocline. However, the seasonal turbocline that corresponds to the daily maxima of the diurnal turbocline can be identified with the depth of the epilimnion and thus with the seasonal thermocline.

The external driving processes influence the content of the TKE in the mixed layer as follows:

(i) The wind input generates TKE by establishing a vertical shear and producing breaking surface waves.
(ii) Heat flow over the free surface into the atmosphere increases the density and thus lowers the buoyancy of the uppermost layers (as long as the temperature stays above the density maximum). The heavy water sinks and mixes itself turbulently with the underlying layers until homogenized conditions are re-established. The TKE of the mixed layers rises.
(iii) Conversely, heat flow from the atmosphere into the lake raises the buoyancy of the uppermost layers. TKE produced by the wind shear must be expended to homogenize the mixed layer by mixing the stable upper layer against gravity into the lower and colder layers.
(iv) The non-uniform heating of the mixed layer by solar radiation stabilizes the water column in a similar way as a heat flow across the free surface, since the absorbed radiative heat is exponentially attenuated with depth. This process also absorbs TKE.
(v) When the mixed layer is deepened, the cold water from below the turbocline must be mixed with the water above, a process that also consume TKE.

On the basis of the two-layer assumption, a positive TKE balance of the processes (i) – (iv) is instantly and completely compensated by lowering the turbocline into deeper layers (entrainment) and thus consuming the abundant TKE via processes (v). In the case of a negative balance, the turbocline is elevated (detrainment) to reduce the depth to which warm surface water has to be mixed down.

In the one-dimensional turbulence model used here, the restriction to such a simplified two-layer system is lifted. Driven by hourly meteorological fields it calculates temperature, turbulent kinetic energy and turbulent dissipation rates on a depth grid with $dz = 0.25$ m and on time steps of four minutes.

3.2 MODEL EQUATIONS

The evolution of all variables is determined by the one-dimensional vertical diffusion equation with time and space dependent vertical diffusion coefficients which are discussed in great detail in Jöhnk and Umlauf (2001) and Hutter and Jöhnk (2004). The equations for temperature, turbulent kinetic energy and its dissipation together with the balance equations for momentum in horizontal direction are solved numerically using a implicit time integration method.

3.2.1 Internal Oscillations

During the periods of stratification, the isotherms nearly never stay at horizontal levels (as in the case of the undisturbed equilibrium) but exhibit periodic fluctuations both in time and space (internal waves and internal oscillations). In a small lake the internal waves are reflected multifold by the nearly omnipresent lateral boundaries, resulting in the quasi-standing patterns of internal oscillations. Although in tropical lakes the temperature differences between epilimnion and hypolimnion are only a few degrees, due to the high absolute values the vertical density gradients are rather large and the stability of the water column is relatively strong. Consequently, the phase velocities of long internal waves are comparable with the values, which are known from the deep lakes of the temperate zones.

Due to the short cross-lake distances it is clear that the periods of free internal oscillations in a small lake should not be larger than a few hours.

We make profit of the fact that under the assumption of flat bottom the spatial dependency of the variables can be separated into a purely vertical and a purely horizontal dependency, respectively (Charney, 1955). The differential equation for the vertical dependency is given by

$$\frac{d^2\hat{Z}}{dz^2} + \frac{N^2(z)}{gh_n}\hat{Z} = 0,$$

which together with the boundary conditions at the surface (rigid lid assumption)

$$\hat{Z} = 0 \quad \text{at} \quad z = 0$$

and at the bottom

$$\hat{Z} = 0 \quad \text{at} \quad z = H$$

form a classical Sturm-Liouville problem for the eigenvalue gh_n, which in the present case is the square of the phase velocity of long internal waves. $N(z)$ is the Brunt-Väisälä frequency defined by

$$N^2(z) = \frac{g}{\rho}\frac{d\rho}{dz},$$

where $\rho(z)$ is the vertical density distribution, which in fresh water lakes is mainly determined by the temperature.

Once c_n is calculated, the system of differential equations governing the horizontal dependency is given by

$$i\omega U + fV + gh_n\frac{\partial Z}{\partial x} = 0,$$

$$i\omega V - fU + gh_n \frac{\partial Z}{\partial y} = 0,$$

$$i\omega Z + \frac{\partial U}{\partial x} + \frac{\partial V}{\partial y} = 0.$$

Here, f is the Coriolis parameter ($f = 0.56 \times 10^{-4}\ \mathrm{s}^{-1}$ for Subhas Sarobar), U, V, Z are the (complex) amplitudes of the components of the horizontal velocity and the vertical displacement of the isothermes, respectively. By solving the systems of equations numerically for the eigenvalues (ω), we get a set of eigenfrequencies, which may be ordered in a chronological way ($\omega_1 < \omega_2 < \omega_3 < ......$). ω_1 is the eigen frequency of the so-called fundamental horizontal mode of oscillation.

The eigen frequencies of the free internal oscillations in the stratified lake are calculated with a proved numerical finite difference model (Bäuerle, 1985).

4. RESULTS AND DISCUSSION

4.1 Simulation of Temperature Stratification

The temperature stratification of the lake, Subhas Sarobar has been simulated for the years 1999 to May, 2004. The model is calibrated by comparing the results of simulations with measurements. The model run is initialised by applying the prescribed atmospheric forcing to a water body at uniform temperature, close to the average air temperature. The total short wave irradiance is derived semi-empirically as described somewhere else (Samal, 2004), all other meteorological parameters are taken from the measured values of a nearby weather station. The results for the year of comparison of simulation with measured data are displayed in the panels of Fig. 2.

The figures show an appreciable good agreement to the measured data. A simple quantification of the goodness of simulation results for individual profiles at different stations within the lake—although the simulation is run only with respect to the deepest point—is made by calculating the mean squared error. The low temperature as observed in the first profile of the measured data as compared to profiles of other date between different stations might be due to differential heating or differential mixing, respectively.

In Fig. 3 the phase velocity of the first baroclinic mode is shown for five years of numerical calculations. Herewith, it is possible to estimate the spectra of internal oscillations.

4.2 INTERNAL OSCILLATION

For the three lowest modes of internal oscillations of the equivalence depth model the dependency of the eigenfrequencies on the phase velocity of long

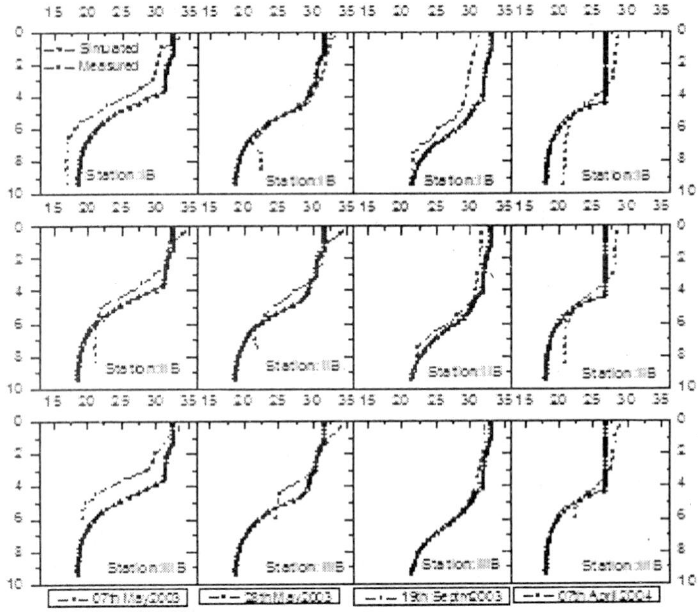

Figure 2: Measured and simulated temperature profile of lake Subhas Sarobar.

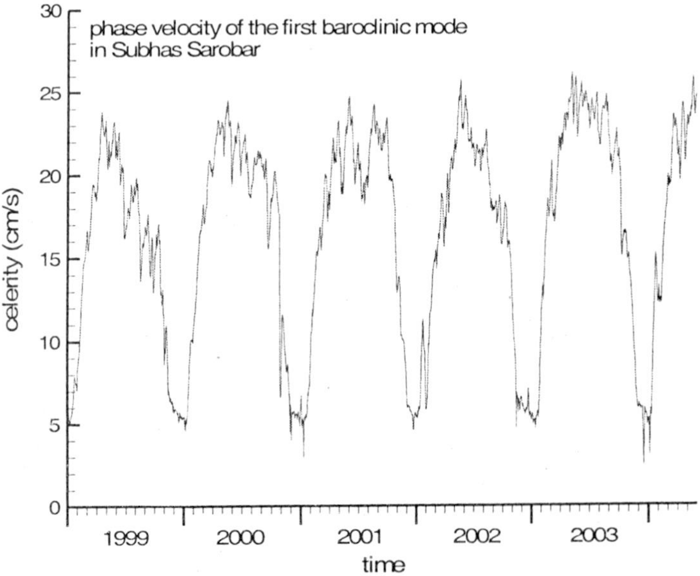

Figure 3: The development of the phase velocity (celerity) of the first vertical mode of long internal waves in Subhas Sarobar from 01.01.1999 to 31.05.2004. The depth of the basin is assumed to be 9.25 m. Stratification is taken from a one-dimensional turbulence model driven by the observed meteorological forces (profiles at 12:00 hours are taken).

internal waves is shown in Fig. 4. The curves follow approximately the formula

$$T_n = \frac{2L^{(n)}}{c} \quad \text{or} \quad \omega_n = \frac{\pi c}{L^{(n)}},$$

where $L^{(n)}$ is a (mode-dependent) length scale, c is the phase velocity of long internal waves (first baroclinic order). The value of 25.0 cm/s corresponds to very strong stratification as observed in summer (Fig. 3). The value of 4.43 cm/s could happen very early in spring or very late in autumn.

We get the length scales $L^{(n)}$ for Subhas Sarobar. by calculating the eigenfrequencies of the model lake for a certain stratification which is represented by a specific value of the phase velocity c_r as a reference (Table 1). Obviously, $L^{(1)}$ and $L^{(2)}$ correspond to the mean length and the mean width of the lake, respectively.

Table 1: The eigenfrequencies and length scales of the four lowest internal seiches in Subhas Sarobar without Earth's rotation. The values of the second column refer to c_r = 4.43 cm/s.

Mode	ω_n ($\times 10^{-4}$ s^{-1})	ω_n/ω_1	$l^{(n)}$ (m)
1	2.25	1.000	618
2	3.62	1.627	380
3	4.34	1.929	320
4	5.04	2.240	276

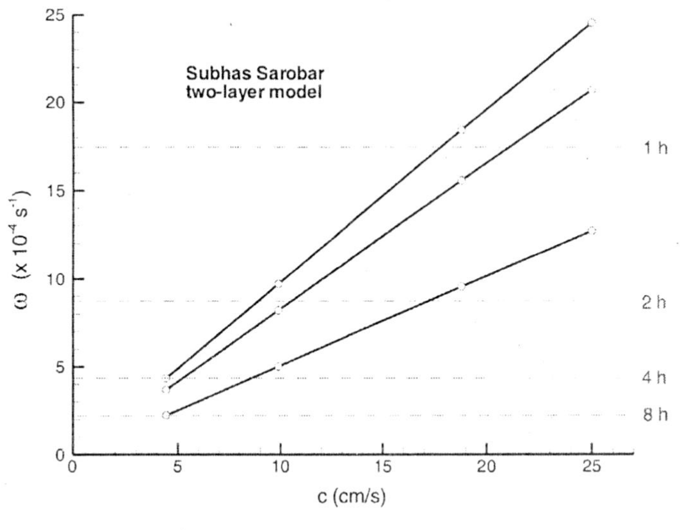

Figure 4: The dependency of the three lowest modes of internal free oscillations in an equivalence-depth model of Subhas Sarobar on the phase velocity of long internal waves. The range of c corresponds to the values of Fig. 3.

5. CONCLUSIONS

The application of a vertical, one-dimensional turbulence model to the tropical lake Subhas Sarobar showed good agreements with temperature measurements at that lake. The calibrated model can now be used to simulate temperature profiles and mixing intensity depending on meteorological forcing. Several applications are possible. Firstly the model can be used to hindcast temperature profiles to fill in measurement gaps in previous years. Secondly it can be used to predict the temperature development in the lake based on actual weather conditions. And lastly it is possible to forecast lake temperature and mixing conditions in the lake under future climate change scenarios. The theoretical findings concerning internal oscillations need observational confirmation. However, based on the results of the numerical calculations it is clear that the main energy of the internal oscillations is at periods of a few hours. With a minimum of temperature sensors it would be possible to throw light on that fascinating subject.

REFERENCES

Bäuerle, E., 1985. Internal free oscillations in the Lake of Geneva. *Annales Geophysicae*, **3:** 199-206.

Charney, J.G., 1955. Generation of oceanic currents by winds. *J. Mar. Res.*, **14:** 477-498.

Hutter, K. and Jöhnk, K.D., 2004. Continuum methods of physical modelling – Continuum mechanics, dimensional analysis, turbulence: Springer-Verlag, Berlin Heidelberg, pp. 635.

Imberger, J., 1998. Flux paths in a stratified lake: Review, Physical processes in lakes and oceans, Coastal and Estuarine studies. *American Geophysical Union*, **54:** 9-10.

Jöhnk, K.D., 2005. Heat Balance of Open Water Bodies. *In:* Lehr, J.H. (ed.), The encyclopedia of Water. Wiley & Sons (in press).

Jöhnk, K.D. and Umlauf, L., 2001. Modelling the metalimentic oxygen minimum in a medium sized alpine lake. *Ecol. Model.*, **136:** 67-80.

Kraus, E.B. and Turner, J.S., 1967. A one-dimensional model of the seasonal thermocline – II. The general theory and its consequences. *Tellus*, **19:** 98-105.

Pawson, L., 2004. Weather Underground, http://www.wunderground.com

Niiler, P.P., 1977. One-dimensional models of the seasonal thermocline. *In:* Goldberg, E.D. (Ed.), The Sea, pp. 97-115.

Saha, T.K., 2000. Final report on "Monitoring of environmental status of Subhas Sarobar, Calcutta and preparation of management action plan" (01.08.1998 - 31.01.2000), Institute of Wetland Management and Ecological Design, Salt Lake, Calcutta - 700 091.

Samal, N.R., Roy, D., Mazumdar, A. and Bose, B., 2004. "Sustenance of aquatic life in a thermally stratified tropical shallow lake", International Conference on Environment and Development organised by Institute of Social Science and West Bengal State of Technical Education, DST, Govt. of West Bengal, India.

Samal, N.R., 2004. Study of morphometry and hydrothermal analysis of a tropical shallow Indian lake, report submitted to DAAD (A/04/06852), Bonn, Germany.

Physical Processes in Large Lakes

Yerubandi R. Rao and Raj C. Murthy

National Water Research Institute
867 Lakeshore Road, PO Box 5050
Burlington, Ontario, Canada L7R 4A6
Ram.Yerubandi@ec.gc.ca

1. INTRODUCTION

The North American Great Lakes lie between 41°N and 49°N and between 76°W and 92°W, on the international boundary between Canada and United States of America. The lakes supply freshwater for over 30 million people along their perimeter and are used for a wide range of activities. With growing coastal population and industrialization, North American Great Lakes are also currently experiencing declining water quality and water supply. These concerns required an international cooperation between the U.S. and Canada in formulation of water quality agreements to protect the water resources. Basic and applied research on the Great Lakes has been conducted by several agencies within Canada and the U.S. to understand physical, biochemical and geological interactions relating to water management issues.

Several large scale cooperative efforts have been carried out in the last three decades which include the International Field Year for the Great Lakes (IFYGL, 1981) on Lake Ontario; the upper Great Lakes reference study (IJC, 1977), which examined the Lakes Huron, Superior and the Georgian Bay; the Lake Erie Binational study (Boyce et al., 1987); Episodic Events Great Lakes Experiment (EEGLE) (Eadie et al., 1997), which examined the sediment plumes in Lake Michigan. In addition several site specific studies are conducted either for delineating the coastal boundary layer on the north shore of Lake Ontario (Rao and Murthy, 2001a) or for providing circulation and diffusion characteristics for siting outfalls in Lake Ontario (Miners et al., 2002). In this article, we will present an overview of some of the circulation features and turbulent exchange processes in a distinct inshore region (10-15 km from the shore) of Lake Ontario (Fig. 1).

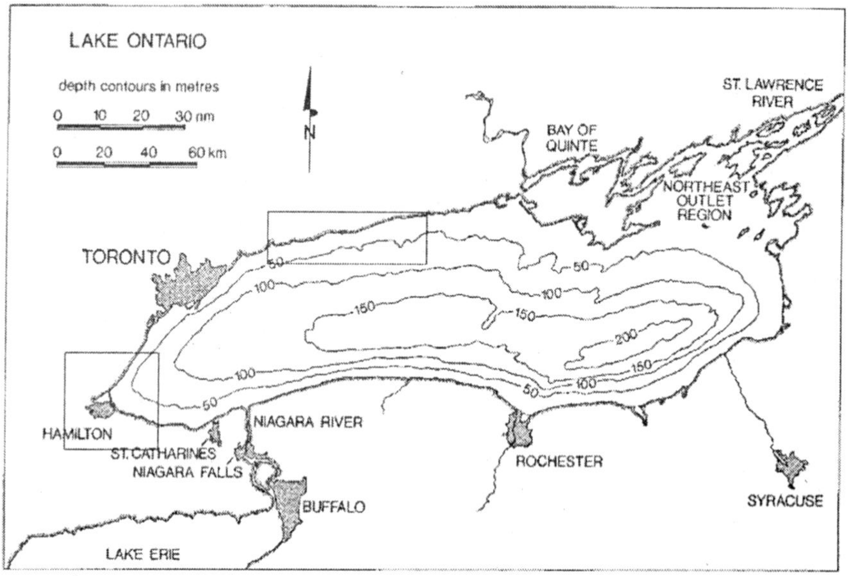

Figure 1: Map of Lake Ontario with bathymetry. The rectangles show the two experimental sites.

1.1 Monitoring of the Coastal Zone

Over the past three decades intensive coastal experiments have been conducted. Earlier work was directed more specifically at waste heat dissipation, and fate of accidental toxic discharge, from large power generating plants along the northshore of Lake Ontario. Recent experiments have dealt with possible effects of municipal waste discharges into the nearshore waters. While instrumentation has improved, the design of the basic coastal experiment has remained unchanged. Along straight shore lines, a series of current meter moorings referred to as a coastal chain are installed in a line perpendicular to the local shore line, at intervals ranging from 500 m to 1 km apart near shore to 2 to 4 km apart at the offshore end of the line. Each current meter records current velocity and water temperature for several months. In recent years the development of the Acoustic Doppler Current Profiler (ADCP) has provided a means of recording accurate current measurements from many vertical levels by a single instrument. Along with current meters, meteorological buoys are moored near the offshore end of the current meter chain to record wind velocity, air temperature, relative humidity, surface water temperature and solar radiation. Fixed temperature profiler (FTP) moorings with several regularly spaced temperature loggers, are often placed at one or more locations to provide detailed vertical temperature data. The basic coastal chain data is frequently operated concurrently with other experimental programmes, such as ongoing water quality monitoring programmes, drifting buoy experiments, or vessel based temperature profiling surveys (Murthy et al., 2000). Flourescent tracers have been used in the past to study coastal dispersion processes. Tracer injection moorings were installed to generate tracer plumes of constant initial concentration. Downstream concentration profiles were measured at several depths using vessel-mounted recording florometers, providing the basis for the determination of mixing lengths and diffusion coefficients. More recently, development of satellite tracked drifting buoys has made it possible to collect fairly long Lagrangian current records indicating water mass transport and dispersion.

2. COASTAL CIRCULATION

Water movements within the coastal zones of the Great Lakes are complex and have particular relevance to water quality concerns. Climatologies of current structure and thermal structure are a prerequisite for many practical problems. Problems encountered, requiring detailed examination of the physical limnology within the coastal zone, include waste disposal through sewage outfalls, large-scale dumping operations, shore erosion and sediment transport, installation of coastal structures, land reclamation, and recreation.

2.1 Thermal Characteristics

The thermal structure and circulation in the Great Lakes generally depends on the season because of the large annual variation of surface fluxes (Boyce et al., 1989). During the spring the thermal bar, for example, acts as a lateral thermal barrier to the transport and dispersion of materials such as pollutants between the nearshore and offshore zones. During the unstratified period (November-June), storm action is the most important forcing, as higher wind speeds and the absence of stratification allow the wind forcing to penetrate deeper into the water column. In summer and fall there is a distinct thermocline in the upper 30 m in most of the lakes which makes them stratified. During this period of stratification, significant wind events will cause upwelling and downwelling of the thermocline along the shore. The scale of the offshore distance over which these events takes place depends on the wind stress and nearshore bathymetry, and is typically of the order of 5-10 km, hence, within the coastal boundary layer (Rao and Murthy, 2001a).

2.2 Coastal Climatology

The time series data provide the basis for a climatological description of the site as a function of wind and other atmospheric variables. Analyses of time series data on currents and temperature exhibit extreme variability on spatial and temporal scales, and hence, statistical concepts traditionally used in the fields of oceanography and meteorology have been adopted to summarize the data. Initially statistical/climatological summaries are prepared to assist in assessing the long-term characteristics of a location. Based largely on the analysis of statistical measures of the flow regimes, the relative frequencies of occurrence and duration of various flow regimes can be identified, and a climatology of the flow regime can be produced for a particular location. An example of the vector plot of currents in western Lake Ontario for a selected period is shown in Fig. 2.

While mean currents largely determine simple dilution rates and transport characteristics, variations in currents due to turbulence and other high frequency perturbations can be very important in dispersing contaminants through mixing and diffusion. The variance in a data record is a measure of these variations. Vector data can be manipulated to find the direction along

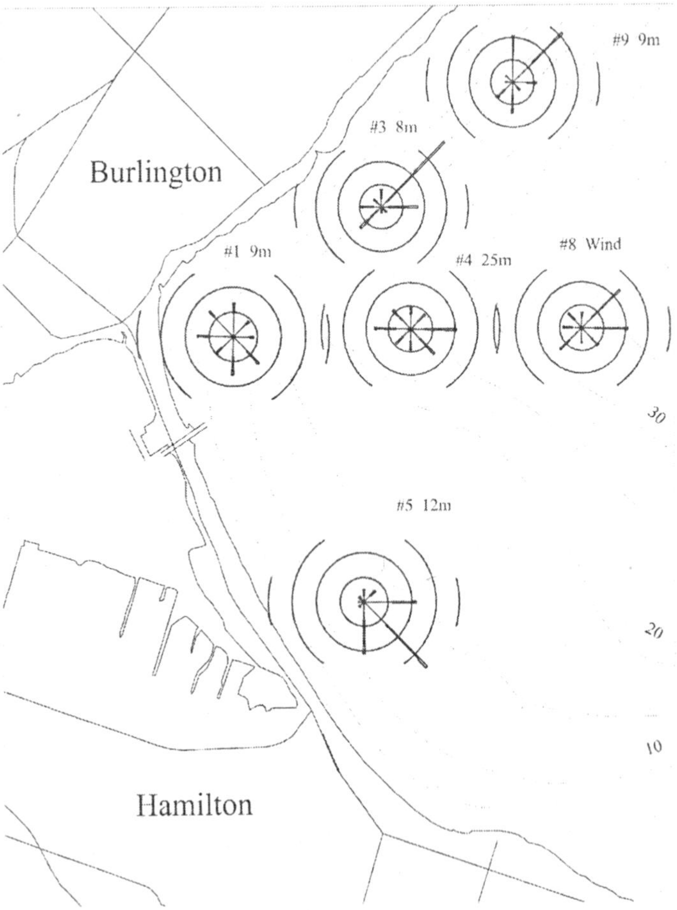

Figure 2: Rose histogram plots of current meter data during summer 1997 in western Lake Ontario.

which the sum of the squares of orthogonal components resolve to an x-axis lying in that direction is a maximum. Such an axis is sometimes called a 'principal axis'. Ellipses with major and minor axes respectively proportional to variance of the flow along and perpendicular to a principal axis were drawn for all current stations to provide an estimate of dispersion at each site. These are shown in Fig. 3 for 1996-97 data. The mean vectors for each segment are also plotted at each station, and, while scales differ, are intended to illustrate that the variance is much greater than the mean current; and hence, is much more important in determining the dispersion.

Individual recorded velocities are the result of motions of several types and spatial scales. Spectral analysis looks at time series data and determines the proportion of the energy (kinetic energy for current or wind data; heat flux for temperature data) attributable to phenomena of different frequencies. For instance, nearshore current data yields high energy values for shore-parallel velocities below 0.01 (~ 4 days) cycles per hour, compared with shore-perpendicular component. The offshore sites show a peak of energy in

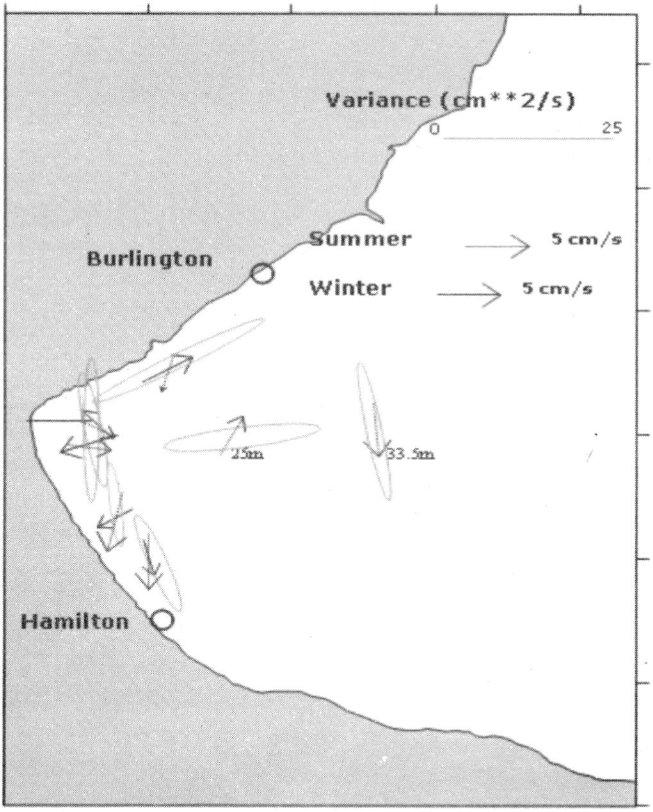

Figure 3: Variance ellipses and mean velocity vectors during 1996-97 in western Lake Ontario.

a narrow band at the inertial frequency. Recently the characteristics of turbulent cross and alongshore momentum exchanges during upwelling-downwelling episodes are obtained in the Lake Ontario (Rao and Murthy, 2001b).

2.3 Coastal Boundary Layer Characteristics

Near the coastal boundaries of large lakes, offshore meandering currents undergo adjustment and move primarily along the boundaries as they approach close to the boundary. The time series of low frequency (filtered, >24 h) flow values $\bar{u}(t)$ and $\bar{v}(t)$ are subtracted from the observed hourly values $u(t)$ and $v(t)$ to obtain the fluctuations u'(t) and $v'(t)$. The variance ($\overline{u'^2}$ and $\overline{v'^2}$) is used as a measure of the magnitude of velocity fluctuations. Turbulence intensity levels in coastal waters can be characterized by root mean square (rms) values as $\sqrt{\overline{u'^2}}$ and $\sqrt{\overline{v'^2}}$. The mean flow kinetic energy, and the fluctuating currents kinetic energy are then simply given as:

$$\{\text{MKE, TKE}\} = \left\{\frac{1}{2}\left(\overline{u^2}+\overline{v^2}\right), \frac{1}{2}\left(\overline{u'^2}+\overline{v'^2}\right)\right\} \tag{1}$$

Figure 4 shows components of kinetic energy (total, mean and fluctuations) as a function of offshore distance. The mean flow kinetic energy dominates within 8-10 km from the shore. Fluctuating kinetic energy or turbulent kinetic energy increases with offshore distance, as near-inertial oscillations become dominant offshore. In summer the MKE increases offshore to a peak at about 3 km from shore then decreases further offshore. Murthy and Dunbar (1981) characterized this flow regime, where total kinetic energy or mean currents increases to a peak as the frictional boundary layer (FBL). Within this zone the currents are influenced by bottom and shore friction. Beyond

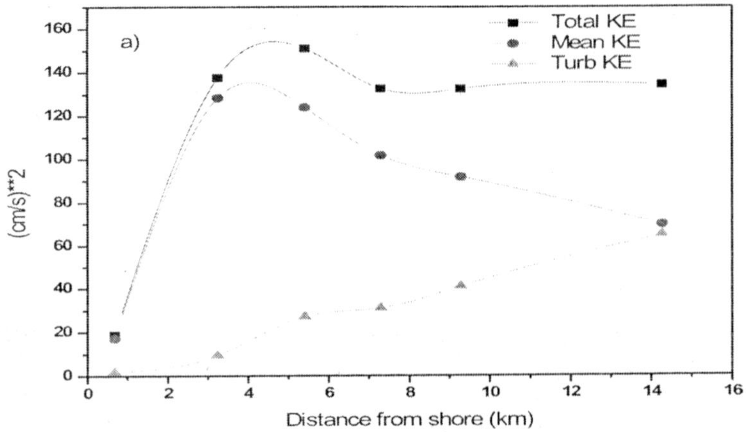

Figure 4: Components of kinetic energy (total, mean and turbulent) with distance from shore in Lake Ontario.

3 km, due to the adjustment of inertial oscillations to shore parallel flow an outer boundary layer develops, known as the inertial boundary layer (IBL). The total (FBL+IBL) forms the coastal boundary layer (CBL). In defining the width of the IBL previous studies used the distance where the inertial oscillations dominate the shore parallel flow. Alternatively, the CBL width can be simply taken as the distance where the TKE contributes maximum to the total kinetic energy. During the summer stratification in Lake Ontario the width of the CBL as determined here was around 10 km, which is consistent with earlier observations (Csanady, 1972).

2.4 Coastal Upwelling/Downwelling

Flow and structure of the coastal boundary layer in the Great Lakes presents a complex scenario during upwelling and downwelling episodes under summer stratified conditions. The theoretical framework which has been created to explain these events comprises two kinds of models. The first model deals with the initial response of the lake to uniform wind stress, and the second type of model deals with the closed nature of the basins wherein transient response is described in terms of internal wave propagation. From the observations we have delineated elements of both theoretical models. In the coastal upwelling zone a near balance exists between wind stress, Coriolis force and internal pressure gradient. However, as the wind subsides two types of waves are established: the Poincare′ wave and the internal Kelvin wave. Poincare′ waves are basin wide response with oscillations in the thermocline across the entire lake with anti-cyclonic phase propagation. On the other hand, internal Kelvin waves are coastally trapped response of the thermocline that progresses cyclonically around the lake. The Rossby radius of deformation which is typically of the order of 3-5 km in the Great Lakes is the e-folding scale for the amplitude of this wave as a function of distance from shore.

An example of coastal upwelling and downwelling circulation along the northshore of Lake Ontario is presented below. During this experimental period temperature variations were dominated by the influence of a few short wind events (Fig. 5). The eastward (westward) wind stress caused thermocline elevation (depression). The upwelling events were characterized by relatively weaker eastward flow, and downwelling events with strong westward currents, with each episode lasting for about 4 to 6 days. The results show inferences to the propagation of internal Kelvin waves due to the thermocline oscillations within the CBL.

2.5 Thermal Bar

A thermal bar is a shore-parallel front which separates descending waters at or near the fresh water temperature of maximum density (4°C) during Spring and Fall seasons. Thermal bars are important because of their influence on

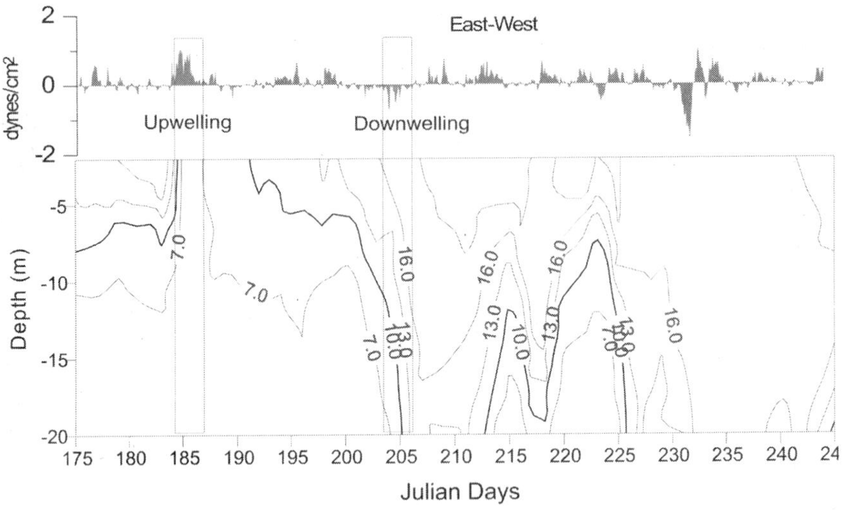

Figure 5: The alongshore wind stress, and temperature profiles in the western Lake Ontario.

mixing, cross-shore exchanges, and the variability of biotic factors. During a thermal bar episode of 17 April to 24 May 1990 in Lake Ontario, the cross-shore exchange coefficients, K_x are nearly constant and consistently smaller than the along-shore counterparts, K_y (Fig. 6). These results suggest that

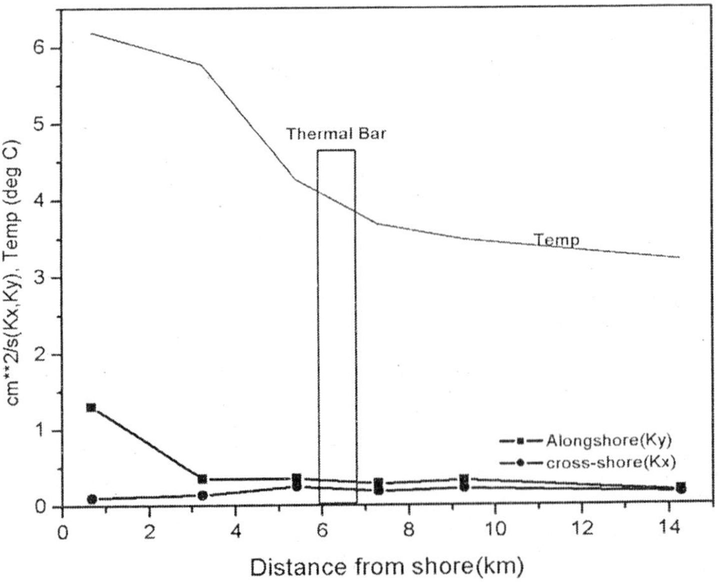

Figure 6: Alongshore and cross-shore exchange coefficients and temperature during the thermal bar episode (17 April to 24 May, 1990).

small-scale horizontal fluctuations and cross-shore turbulent momentum exchanges are severely inhibited in the spring during the thermal bar.

3. HORIZONTAL DIFFUSION

Horizontal exchange of heat, energy, and momentum occur at scales generally much larger than vertical exchanges, given the large horizontal lengths of natural basins compared to their depths. This has prompted many researchers to de-couple horizontal transfers from vertical ones, even though horizontal and vertical processes interact. Numerous dye studies and tracer experiments conducted to determine the diffusion characteristics of the water column also show that the vertical exchange coefficient is much smaller than the horizontal counterparts. Here, we illustrate the nature of horizontal diffusion characteristics in the Great Lakes from a set of carefully conducted experiments (a) Satellite tracked Lagrangian drifters, and (b) using point current meters or acoustic doppler current profilers.

3.1 Lagrangian Drifter Experiments

Drifters have been used to study particle dispersion by oceanic turbulence since the early work of Stommel (1949). Since that time drifter trajectories have been obtained for a wide range of length and time scales. It quickly became apparent that eddy diffusion in oceans and lakes is not Fickian. Taylor (1921) showed that, in a stationary homogeneous turbulence, particle dispersion is related to Lagrangian integral time scales. Here we report drifter trajectories in an experiment conducted in Lake Ontario for two different flow regimes (Figs 7a and 7b). The Lagrangian integral time scale (T_i^L) and length scale (L_i^L) are the time and distance over which the drifter motion remains correlated are given by

$$T_i^L = \int_0^T R_{ii}^L(\tau)d\tau \quad \text{and} \quad L_i^L = \sqrt{\langle u_i'^2 \rangle}\int_0^T R_{ii}^L(\tau)d\tau \tag{2}$$

where $R_{ii}^L(\tau)$ is the auto correlation function defined as

$$R_{ii}^L(\tau) = \frac{1}{T}\frac{\int_0^{T-\tau} u_i'(t)u_i'(t+\tau)dt}{< u_i'^2 >} \tag{3}$$

where u' is the residual velocity defined by $u' = u - <u>$, $<u>$ denotes average over time. Further, it was observed that because of low frequency motions, Lagrangian integral time and length scales are generally time dependent and do not approach a constant limit. Most of the individual autocorrelation functions oscillate and have significant lobes, which underestimate the integral time-scale as they are integrated over the duration

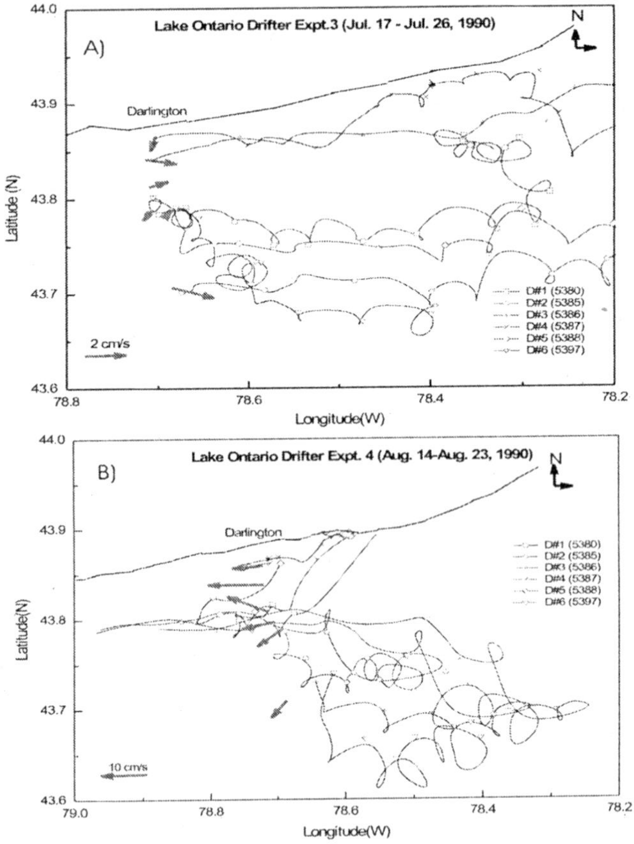

Figure 7: Trajectories of drifters deployed along the northshore of Lake Ontario during two experiments in 1990.

of the time series. To avoid this, we follow the usual practice of integrating from zero to the time of first zero crossing.

Following earlier discussion it is assumed that the drifter velocity fluctuations are homogeneous and stationary as a first order approximation, and hence we can write the mean squared dispersion due to single-particle motion as

$$\left\langle x_i'^2 \right\rangle = 2\left\langle u_i'^2 \right\rangle \int_0^t (t-\tau) R_{ii}^{L}(\tau) d\tau \tag{4}$$

When diffusion time elapses beyond some lag time t_ℓ (Lagrangian correlation time scale), $R^{L}_{ii}(\tau)$ will drop to zero. Physically t_ℓ is the decay time scale of those eddies which contribute to diffusion. Therefore, for large time scales $t > t_\ell$ the horizontal eddy exchange coefficient is given by

$$K^{L}_{i} = \left\langle u_i'^2 \right\rangle T_i^{L} \tag{5}$$

3.2 Eulerian Experiments

In the previous section, statistics in which drifter velocity was a function of its Lagrangian coordinates and times were considered. In a stationary and homogeneous turbulence, the Lagrangian variance $\langle u_l'^2 \rangle$ can be assumed to be equivalent to Eulerian variance $\langle u_e'^2 \rangle$. Hay and Pasquill (1959) also pointed out that the essential difference between Eulerian and Lagrangian velocities is that, at a fixed point, velocity fluctuations appear to move rather quickly, as turbulent eddies are advected past the instrument. They have shown that the Lagrangian correlation function $R_{ii}^L(\tau)$ and the Eulerian counterpart $R_e(\tau)$ have similar shape but differ only by a factor β which is greater than unity. $R_e(\tau) = R_{ii}^L(\beta\tau)$. Introducing these assumptions, the horizontal exchange coefficient in terms of Eulerian statistics can be written as

$$K_e = \beta \langle u_e'^2 \rangle T_e \tag{6}$$

where T_e is the Eulerian integral time scale and $\beta = 1.4$ (Schott and Quadfasel, 1979).

Table 1 presents the horizontal exchange coefficients obtained by Eulerian and Lagrangian measurements during two experiments. In the first experiment during which significant upwelling occurred (see Fig. 7a), the statistics show that alongshore exchange coefficients (K_x) were slightly higher than cross-shore components (K_y) in the first 5.5 km from the shore, i.e. in the FBL. The cross-shore components reached a peak at around 6-7 km from shore and remained steady outside the CBL. These results indicate that momentum transfers occur in the longshore direction in the FBL and cross-shore transfers may dominate in the IBL. Although the magnitude of alongshore Lagrangian eddy coefficients were higher than Eulerian values, they show a peak nearly at the same distance. The cross-shore exchange coefficients in the surface levels were lesser than sub-surface values in the IBL. During the second experiment that is favoured by downwelling circulation (Fig. 7b) the alongshore components were higher in the CBL, and outside the CBL the cross-shore exchanges were dominant. The turbulent momentum exchanges were rather small in the FBL, but significantly increased in the IBL. The exchange coefficients obtained at 3.5 m depth from drifter trajectories have also shown higher alongshore values within the CBL. Both alongshore and cross-shore values increased rapidly to high values outside the CBL with increasing Lagrangian time scales. As observed in the upwelling case, the cross-shore exchange coefficients from Lagrangian measurements were smaller compared to Eulerian coefficients.

Table 1: Alongshore (K_x) and cross-shore (K_y) eddy diffusivities from Eulerian and Lagrangian measurements during upwelling and downwelling cycles (subscripts L indicate Lagrangian, and E indicates Eulerian)

Distance from shore (km)	$K_{x(E)} 10^5\ cm^2 s^{-1}$	$K_{y(E)} 10^5\ cm^2 s^{-1}$	*Drifter bin*	$K_{x(L)} 10^5\ cm^2 s^{-1}$	$K_{y(L)} 10^5\ cm^2 s^{-1}$
			Upwelling		
0.68	0.277	0.864	1	7.13	5.92
3.24	10.36	6.173	2	68.8	10.2
5.42	26.98	16.78	3	15.5	2.02
7.30	16.62	20.38	4	13.4	2.28
9.28	19.26	20.42	5	17.5	6.43
14.2	24.30	20.33	6	27.8	2.75
			Downwelling		
0.68	1.010	0.652	1	6.51	0.35
3.24	7.161	4.197	2	23.1	3.43
5.42	30.97	21.55	3	14.7	4.10
7.30	31.16	28.42	4	83.2	11.6
9.28	37.03	41.57	5	201.2	31.9
14.2	60.09	72.37	6	214.3	48.8

4. CONCLUSIONS

This paper presents an overview of wind-driven coastal circulation and turbulent diffusion processes in the aquatic environment. The results in this article are drawn from large-scale field experiments conducted in the Great Lakes. Although the Great Lakes are easier to study than oceans because they are smaller and do not have salinity effects and tides, still many of the physical phenomena, particularly the circulation forced by large scale winds is similar to that of associated with coastal oceans or inland seas. The coastal circulation and dispersion properties of receiving waters are studied primarily to control hygienic and aesthetic harm to the environment, and to control possible ecological disturbances and modifications. There are several conceptual and theoretical models concerning the dynamics of large-lake systems. However, it is difficult to identify the relative effects of several physical mechanisms to arrive at acceptable predictive hydrodynamic and water quality models. Therefore, large-lake research places heavy emphasis on carefully designed model-driven experiments to collect synoptic and time-series data on currents, thermal structure and diffusion characteristics. These data are used to develop and validate water quality models to simulate observed biogeochemical parameters and processes in large-lake systems and coastal oceans.

REFERENCES

Boyce, F.M., Donelan, D.A., Hamblin, P.F., Murthy, C.R. and Simons, T.J., 1989. Thermal structure and circulation in Great Lakes., *Atmos-Ocean*, **27(4):** 607-642.

Csanady, G.T., 1972. The Coastal Boundary Layer in Lake Ontario: Part II. The Summer-Fall Regime. *J. Phys. Oceanogr.*, **2:** 168-176.

Eadie, B.J., Schwab, D.J., Leshkevich, G.L., Johengen, T.H., Assel, R.A., Holland, R.E., Hawley, N., Lansing, M.B., Lavrentyev, P., Miller, G.S., Morehead, N.R., Robbins, J.A. and Vanhoof, P.L., 1996. Anatomy of a recurrent episodic event: A winter-spring plume in southern Lake Michigan, *EOS. Transactions of the American Geophysical Union*, **77:** 337-338.

Hay, J.S. and Pasquill, F., 1959. Diffusion from a continuous source in relation to spectrum and scale of turbulence. *Adv. in Geophys.*, **6:** 345-365.

International Joint Commission (IJC), 1988. Revised Great Lakes Water Quality Agreement of 1978 as amended by protocol signed November 18, 1987, Washington D.C. and Ottawa.

IFYGL, 1981. The International Field Year on Great Lakes. *In*: E.J. Aubert and Richards, T.L. (ed.), NOAA/GLERL. Ann. Arbor, Michigan.

Lumley, L. and Panofsky, H.A., 1964. The structure of atmospheric turbulence. Interscience, New York, 239 pp.

Murthy, C.R. and Dunbar, D.S., 1981. Structure of flow within the coastal boundary layer of the Great lakes. *J. Phys. Oceanogr.*, **11:** 1567-1577.

Miners, K.C., Chiocchio, F., Rao, Y.R., Pal, B. and Murthy, C.R., 2002. Physical Processes in Western Lake Ontario for Sustainable Water Use. Sci. Rep. No. 02-176, National Water Research Institute, Burlington, Canada, pp. 176.

Rao, Y.R. and Murthy, C.R., 2001a. Coastal Boundary Layer Characteristics during Summer Stratification in Lake Ontario. *J. Phys. Oceanogr.*, **31(4):** 1088-1104.

Rao, Y.R. and Murthy, C.R., 2001b. Nearshore currents and turbulent exchange characteristics during upwelling and downwelling events in Lake Ontario. *J. Geophys. Res.*, **106(C2):** 2667-2678.

Schott, F. and Quadfasel, D., 1979. Lagrangian and Eulerian measurements of horizontal mixing in the Baltic. *Tellus*, **31:** 138-144.

Stommel, H., 1949. Trajectories of small bodies sinking through the convection cells, *J. Mar. Res.*, **8:** 199-225.

Taylor, G.I., 1921. Diffusion by continuous movements. *Proc. London Math Soc.*, **20:** 196-212.

Modelling Coastal Ecology

Girija Jayaraman and Anumeha Dube

Centre for Atmospheric Sciences
Indian Institute of Technology Delhi, New Delhi - 110016
jgirija@cas.iitd.ernet.in, anu_dube@hotmail.com

1. INTRODUCTION

Lagoons, estuaries and other semi-enclosed coastal systems are usually high biological production areas. The physical and chemical dynamics and ecology of shallow estuarine areas are strongly influenced by the runoff of freshwater from the land and the exchange of water with the adjacent open sea. The freshwater input influences estuarine hydrography by creating salinity gradients and stratification and assures large transport of silt, organic material and inorganic nutrients to the estuaries. The open marine areas impose large scale physical and chemical forcing on the estuarine ecosystem due to tide and wind generated water exchange.

Understanding community ecology and ecosystem function of coastal areas is essential for assessing the effects of coastal development and framing coastal management strategies. Biomonitoring or observing how aquatic plants and animals respond to anthropogenic changes can be a significant indicator of the level of efficiency of restoration strategies. Biological monitoring compliments physico-chemical monitoring since the physical and chemical data measure the concentration of pollutants at an instant but biological information can reveal the history or the past effect of any discharge of pollutants on the organisms. Hence, studies on coastal ecology are directed towards finding how organisms in the coastal waters interact with each other and with the physical, chemical, and geological processes in the coastal areas.

In order to obtain solutions to environmental problems associated with coastal areas, we should support a holistic approach to coastal environmental management. Traditional monitoring programmes have to be strengthened, laboratory experiments need to be conducted and mathematical models have to be formulated to evaluate the present state of the system as well as the effectiveness of past and proposed coastal management strategies. In conjunction with monitoring programmes and laboratory experiments, there is a need to formulate quantitative models to predict, to guide assessment and to direct intervention.

1.1 Modelling Approach

During the past few years there is a growing awareness to use modelling principles to study the coastal transport processes-hydrodynamics, sediment transport, and biological transport. Modelling should not be thought of as a theoretical exercise; it is a rational way to put together all the available data and knowledge about the coastal system to get insight into the working of the coastal processes and make quantitative predictions. The strength of mathematical modelling lies in the freedom to meddle with the system i.e., to incorporate alternative mechanisms into the model and investigate different scenarios. This flexibility allows the modeller to test and develop hypothesis regarding the governing mechanism, forecast the time evolution of the system and ascertain the impact of alternative management strategies. Models, furthermore, provide a unifying framework for thinking about the interplay between the various factors that govern a system. There has been more advancement in the formulation and numerical treatment of formulated equations of dynamic models whose purpose is to understand coastal dynamics —tides, storm surges and currents. These models, after validation, are used effectively in coastal management, especially in technical planning, flood control and estimating level of inundations.

1.2 Ecological Models for Coastal/Marine Environment

The circulation/dynamic models are fundamental for any further understanding of the biological processes since the circulatory pattern regulates the supply of dissolved inorganic nutrients to the surface layers where the biomass has to reside to utilize the sun's energy. There has been a lot of advancement in modelling the dynamics of the flow in the lagoons, estuaries and coastal areas but modelling the ecology of these natural aquatic environment is relatively new. The main difficulty is because (i) unlike the dynamical models which have a well-defined starting level in the form of Navier-Stokes equations, the ecological models do not have a prescribed starting point and (ii) the variance associated with any set of data is usually very much greater than the mean and hence testing the theories is not easy (Steele, 1975). Early models of marine/aquatic ecology were based on population dynamics. While these time dependent models in the form of dynamical systems have been very helpful in understanding biological processes, they did not explicitly include the effects of advection and diffusion on the ecological system. But with the advancement of computational techniques and technological sophistication in recording the data at several spatial locations, a holistic approach leading to spatiotemporal models is getting popular (Pinazo et al.,

2004; Griffin et al., 2001; Flindth et al., 1999; Ryabchenko et al., 1998; Oguz et al., 1996). The objective of the mathematical modelling of ecological problems is quantification of the interaction within and between species and their inorganic environment and the investigation of the temporal/spatial variation of groups of individual of various species.

This article will deal only with coastal ecological models, starting from the simplest ones to their current status. The basic equations, the intra-species interaction are all explained in Section 2. Every aquatic ecosystem is a unique natural feature, and hence, it is impossible to make generalized statements. The models can be evaluated only by applying them to a case study where necessary data is available. With this in mind, as an illustration, a case study corresponding to the largest lagoon in India—the Chilika lagoon is discussed in Section 3. Finally, the limitations of the modelling approach are discussed in Section 4.

2. THEORY BEHIND COASTAL ECOLOGICAL MODELS

Modelling coastal ecology is mostly based on following the food chain: mechanisms by which the phytoplankton consumes the nutrients, the zooplankton feeds on the phytoplankton, the carnivores on the zooplankton, the big fish on the carnivores etc. Figure 1 gives a schematic diagram of the food web which is followed while formulating a mathematical model. The models are made on the basis of these compartments, which may be biotic (e.g. zooplankton, whales) or abiotic (e.g. nitrate, detritus, dissolved organic carbon (DOC)). The procedure followed is to build a model by first defining

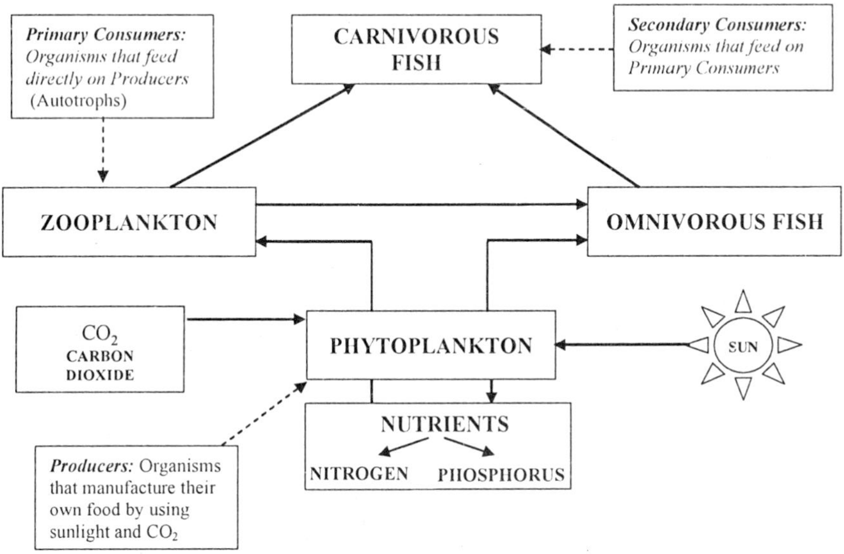

Figure 1: Aquatic food web.

the compartments and then specify the equations governing the transfer of material between compartments.

Mathematical models for marine/coastal ecology were first developed in the 1940s (Riley, 1946, 1947) to understand the seasonal patterns in plankton abundance keeping in mind that large changes in plankton populations can have far reaching consequences in terms of food supply or climate change. The first few models which were inspired by these early models were time dependent first order differential equations (Evans and Parslow, 1985; Franks and Chen, 1996; Fasham et al., 1990; Fasham et al., 1993) whose formulation required knowledge of (1) growth rate of phytoplankton as a function of irradiation and nutrient limitation, (2) loss rates for phytoplankton from their mortality, sinking, diffusing and grazing by zooplankton, (3) changes in nutrient concentrations due to the uptake by planktons, regeneration, mixing due to turbulent waters which bring the bottom nutrition to the surface etc., (4) growth rate of zooplankton as a function of phytoplankton concentration, nutrition and its own efficiency in food assimilation, and (5) loss rates of zooplankton due to natural mortality or grazing by higher trophic levels. A simple three compartment NPZ (Nutrient, Phytoplankton, Zooplankton) model (Steele and Henderson, 1981) ignoring spatial gradients, can be represented by three coupled ordinary differential equations as follows:

$$\frac{dP}{dt} = \text{Production of phytoplankton} - \text{Grazing by zooplankton} - \text{Mortality losses} \quad (1)$$

$$\frac{dN}{dt} = -\text{Production of phytoplankton} - \text{Grazing by zooplankton} - \text{Mortality losses} \quad (2)$$

$$\frac{dZ}{dt} = \text{Production of phytoplankton} - \text{Grazing by carnivores} - \text{Mortality losses} \quad (3)$$

Including more number of species in the model need not necessarily make the model more realistic since it may make the accuracy of the model worse. Few species model can be more instructive since one can ensure taking into account the major functional responses or interactions of the species. One of the best models so far is that of Fasham (1990) who considered a nitrogen-based model of plankton dynamics in the oceanic mixed layer. The model studied the annual cycles of phytoplankton using seven compartments (Phytoplankton, Zooplankton, Bacteria, Nitrate, Ammonium, Dissolved organic nitrogen and Detritus).

The distribution of the species changes not only with time but also with space. Studying the spatial heterogeneity or plankton patchiness demand more sophistication in the models. Differences in spatial mobility, bioconvection and gyrotaxis (Pedley and Kessler, 1992) may also be responsible for these spatial structures. Some of the initial studies on modelling coastal ecology concentrated on the vertical distribution and hence the diffusion of the biological variables like nutrients, phytoplankton, zooplankton, fishes etc. (Tett, 1981). The objective was to describe nutrient (dissolved and cellular) and phytoplankton distribution in the vertical direction due to turbulent diffusive processes. The next step in modelling was to include the effect of circulation and hence introduce physico-biological models. The wide range of scales corresponding to space and time make implementation of these models (Pinazo et al., 2004; Griffin et al., 2001; Flindt et al., 1999; Ryabchenko et al., 1998; Oguz et al., 1996) quite a formidable task.

We restrict our discussion only to time dependent models which involve rate equations and shall describe the specific functional forms used in equations (1-3) along with the case study to be discussed in Section 3.

3. CHILIKA LAGOON - A CASE STUDY

Chilika Lagoon (19°28′ N to 19°54′ N and 85°06′ E to 85°35′ E) is the largest brackish water lagoon with estuarine character (Fig. 2).

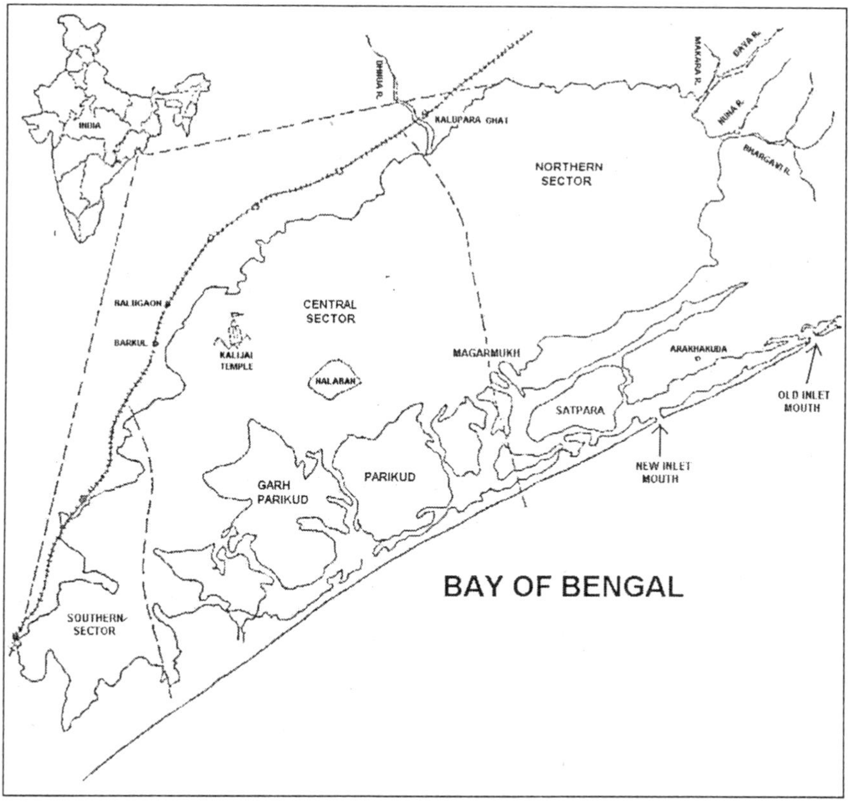

Figure 2: Map of Chilika showing different sectors.

Interest in detailed analysis of the ecology of the lagoon and the various factors affecting it is due to the opening of the new mouth on 23 September 2000 to resolve the threat to its environment from various factors—eutrophication, weed proliferation, siltation, industrial pollution and depletion of bioresources.

In an earlier study (Jayaraman et al., 2006), the dynamics and salinity distribution of Chilika was modelled. For a holistic study of the processes affecting the lagoon, it is important to model the seasonal variations in the planktonic species. With this objective, in this section, a purely ecological minimal model for simulating the annual plankton dynamics of the lagoon is presented. The model tries to study the ecology through a three-compartment (NPZ) formulation resulting in dynamical system of equations. This model is solved for different sectors of the lagoon which have distinct characteristics in terms of depth, light penetration, nutrients, salinity and hence planktonic growth.

3.1 Mathematical Formulation

A three-compartment (NPZ) model for nutrients (N), phytoplankton (P) and zooplankton (Z) is proposed here. The specific governing equations for the three variables N, P, Z considered in this study are:

$$\frac{dN}{dt} = -\left[\frac{\bar{\alpha}\ N}{K_{\mathrm{N}} + N} - r\right] P + \frac{m}{D}\ N_0(t) \tag{4}$$

$$\frac{dP}{dt} = \left[\frac{\bar{\alpha}\ N}{K_{\mathrm{N}} + N} - r\right] P - \frac{c\ (P - P_0)\ Z}{(K_{\mathrm{Z}} + P - P_0)} \tag{5}$$

$$\frac{dZ}{dt} = \frac{e \times c \times (P - P_0) \times Z}{(K_{\mathrm{Z}} + P - P_0)} - g \times\ Z \tag{6}$$

where the concentrations of nutrients, phytoplankton and zooplankton are measured in $\mathrm{mMNm^{-3}}$, t is time, $\bar{\alpha}$ ($\mathrm{day^{-1}}$) is the light limited growth rate of phytoplankton, and r ($\mathrm{day^{-1}}$) is the respiratory loss rate of phytoplankton. K_{N} ($\mathrm{mMNm^{-3}}$) and K_{Z} ($\mathrm{mMNm^{-3}}$) are the half saturation coefficients for nutrient uptake and zooplankton grazing respectively, c ($\mathrm{day^{-1}}$) is the grazing rate and P_0 ($\mathrm{mMNm^{-3}}$) is the grazing threshold. $N_0(t)$ is the source of nutrients, m ($\mathrm{day^{-1}}$) is the vertical diffusion rate and D is the depth in metres. The e is the grazing efficiency and g ($\mathrm{day^{-1}}$) is the loss of zooplankton to carnivores. The values of the parameters fixed for the validation of the model values with the observed data are given in Dube and Jayaraman (2006). A detailed term-by-term explanation of the equations (2-4) is given in Evans and Parslow (1985).

3.2 Numerical Experiments and Prediction of Range of values for the Parameters

Numerical experiments are performed to examine the relative importance of different parameters involved in the model. The time step used in the numerical integration of the model is 15 minutes. Though most parametric values are taken from observed data/literature, ranges of some of the unknown parameters are found using the stability criteria for dynamical systems (Dube and Jayaraman, 2006).

The results obtained from the above defined model are given in Figs 3 to 5. These results have been validated against the data given by Adhikary and Sahu (1992), which are the only time series observations available. However the units of measurement used by them are number of phytoplankton per litre so only a qualitative validation of the results is possible.

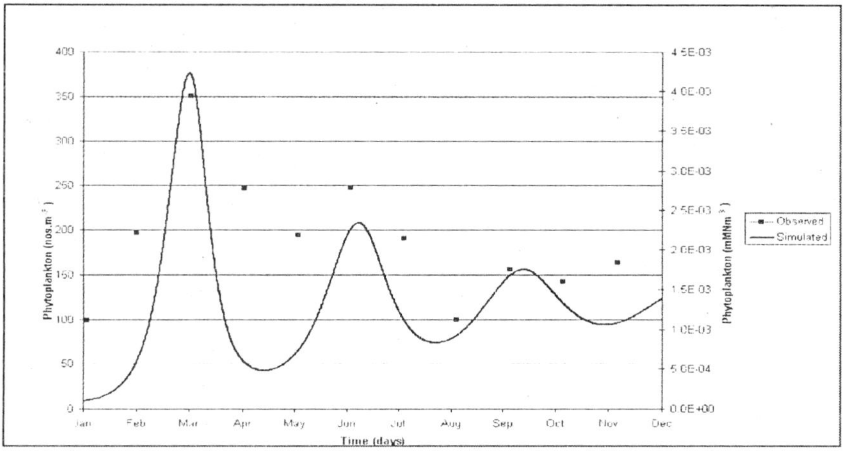

Figure 3: Annual distribution of phytoplankton in northern sector

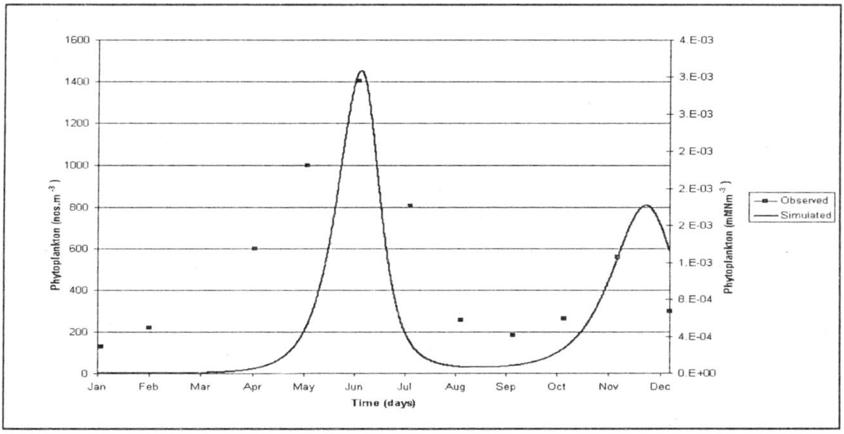

Figure 4: Annual distribution of phytoplankton in central sector

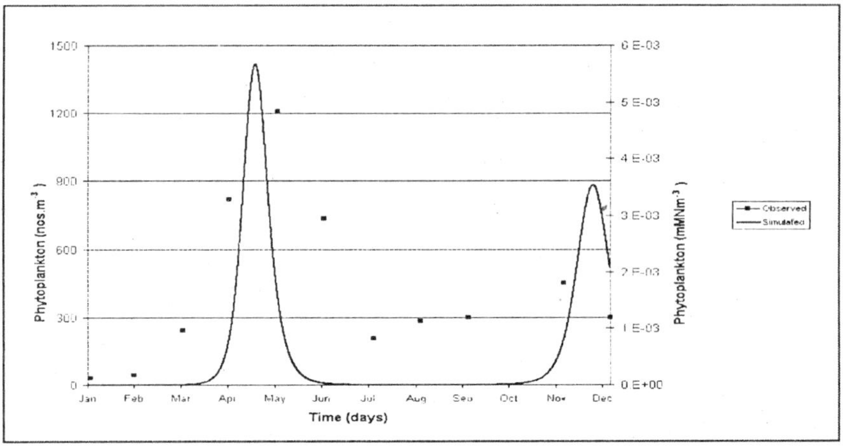

Figure 5: Annual distribution of phytoplankton in southern sector

3.3 Conclusions of the Chilika Model

Numerical experiments and sensitivity tests with different parameters show that the main parameter affecting the productivity is the growth rate, which depends on light and nutrients. It is also found that the time of maximum planktonic growth is mostly dependent upon the parameter values within the growth rate term. The major peaks of phytoplankton in all the sectors of Chilika, as seen in the data (Adhikary and Sahu, 1992), have been well reproduced by the model. Though we have validated for only one year, we have also verified for different years (1987, 1998, 1999, 2000 and 2001) using the discrete data we have from other authors.

For a holistic and complete understanding of the influence of physical processes, freshwater discharges and tidal forces on the seasonality of phytoplankton in Chilika lagoon, a physico-biological model linking an earlier study (Jayaraman et al., 2006) with the present analysis needs to be formulated.

4. LIMITATIONS OF MODELS

Ecological models have to be supported by reliable observations before we rush to use them for predictions or coastal management. It must be remembered that the real problem is far more complicated, number of state variables are far too many, and the model cannot include all the interactions exhaustively. A lot of tuning goes into a model to validate it against field data but one cannot say with certainty if the model will be fully reliable for prediction purposes. The increased power of computing and the improved biological data supported by satellite observations collectively promise and encourage the modellers to take up more challenging problems which include most of the spatio-temporal characteristics of the coastal transport processes.

REFERENCES

Adhikary, S.P. and Sahu, J.K., 1992. Distribution and Seasonal abundance of Algal Forms in Chilika Lake. *Japanese Journal of Limnology*, **53(3):** 197-205.

Dube, A. and Jayaraman, G., 2006. Modelling the seasonal variability of planktons in Chilika lagoon, east coast of India. (Communicated), *Journal of Applied Ecology*.

Evans, G.T. and Parslow, J.S., 1985. A model of annual plankton cycle. *Biological Oceanography*, **3:** 327-347.

Fasham, M.J.R., Ducklow, H.W. and Mckelvic, S.M., 1990. A nitrogen based model of plankton dynamics in the oceanic mixed layer. *Journal of Marine Research*, **48:** 591-639.

Fasham, M.J.R., Sarmiento, J.L., Slater, R.D., Ducklow, H.W. and Williams, R., 1993. Ecosystem behavior at Bermuda station "S" and ocean weather station "India": A general circulation model and observational analysis. *Global Biogeochemical Cycles*, **7:** 379-415.

Flindt, M.R., Pardal, M.A., Lillebo, A.I., Martins, I. and Marques, J.C., 1999. Nutrient cycling and plant dynamics in estuaries: A brief review. *Acta. Oecol.*, **20(4):** 237-248

Franks, P.J.S. and Chen, C., 1996. Plankton production in Tidal Fronts: A Model of Georges Bank in Summer. *Journal of Marine Research*, **54:** 631-651.

Griffin, A.L., Michael, H. and Hamilton, D.P., 2001. Modelling the Impact of Zooplankton Grazing on Phytoplankton Biomass during a Dinoflagellate Bloom in the Swan River Estuary, Western Australia. *Ecological Engineering*, **16:** 373-394.

Jayaraman, G., Rao, A.D., Dube, A. and Mohanty, P.K., 2006. Numerical simulation of circulation and salinity structure in Chilika Lagoon. *Journal of Coastal Research*. (In Press).

Oguz, T., Ducklow, H., Malanotte-Rizzoli, P., Tugrul, S., Nezlin, P.N. and Unluata, U., 1996. Simulation of annual plankton productivity cycle in the Black Sea by a one-dimensional physical-biological model. *Journal of Geophysical Research*, **101(7):** 16585-16600.

Pedley, T.J. and Kessler, J.O., 1992. Hydrodynamic Phenomena in Suspensions of Swimming Microorganisms. *Annual Review of Fluid Mechanics*, **24:** 313-358.

Pinazo, C., Bujan, S., Douillet, P., Fichez, R.C., Grenz, C. and Maurin, A., 2004. Impact of wind and freshwater inputs on phytoplankton biomass in the coral reef lagoon of New Caledonia during the summer cyclonic period: a coupled three-dimensional biogeochemical modelling approach. *Coral Reefs*, **23(2):** 281-296.

Riley, G.A., 1947. A theoretical analysis of the zoo-plankton population on Georges Bank. *J. Mar. Res.*, **6:** 104-113.

Riley, G.A., 1946. Factors controlling phytoplankton populations on Georges Bank. *J. Mar. Res.*, **6:** 54-73.

Ryabchenko, V.A., Gorchakov, V.A. and Fasham, M.J.R., 1998. Seasonal dynamics and biological productivity in the Arabian Sea euphotic zone as simulated by a three-dimensional ecosystem model. *Global Biogeochemical Cycles*, **12:** 501-530.

Steele, J.H., 1975. Biological Modelling, Chapter 10: Modelling of Marine Systems. J.C.J. Nihoul (ed.), Elsevier, pp. 208-216.

Steele, J.H. and Henderson, E.W., 1981. A Simple Plankton Model. *The American Naturalist*, **117(5):** 676-691.

Tett, P., 1981. Modelling phytoplankton production at shelf-sea fronts. *Philosophical Transactions of the Royal Society of London*, **A302:** 605-615.

Adaptation to Salinity Change Induced by Sea-Level Rise in Hinuma Lake, Japan

Hisamichi Nobuoka and Nobuo Mimura

Ibaraki University, Ayukawa 6-9 A 204, Hitachi, Ibaraki, Japan
nobuoka@mx.ibaraki.ac.jp

1. INTRODUCTION

Brackish water lake is home of peculiar organisms, as its salinity is middle of those of sea and fresh water. Its salinity changes sensitively with sea level, river flow rate and lake topography. Hinuma Lake is one of brackish water lakes in Japan and connects with the Pacific Ocean through a tributary and a major river. Therefore the inflow process of salt water to the lake is very complex. The purposes of this study are to estimate impact of sea-level rise on the lake salinity and to examine adaptation options to preserve ecosystem against the impact. To this end, we carried out a long-term observation on the inflow of sea water and seasonal change of salinity to understand the mechanism of sea water intrusion and to verify the numerical simulation model. The influence of topographic changes on salinity is also investigated for the last five decades.

2. FEATURE OF HINUMA LAKE

Hinuma Lake is located about 80 km north of Tokyo in Japan. Figure 1 shows the topography of Hinuma river basin. This lake connects with Pacific Ocean through a tributary Hinuma River (8 km), and the main stream, Naka River (0.5 km). The flow rate of Naka River is ten times as large as that of Hinuma River; which means that the salt water intrusion to the lake is mainly governed by Naka River. The water depth profile of Hinuma River is shown by Fig. 2 which is strongly irregular. The lake is shallow, 2 m in average, and shallower around lake head and mouth.

Fresh water clam, Corbicula japonica Primes, is surviving in the Hinuma basin. The number of the clam has, however, been decreasing for the last

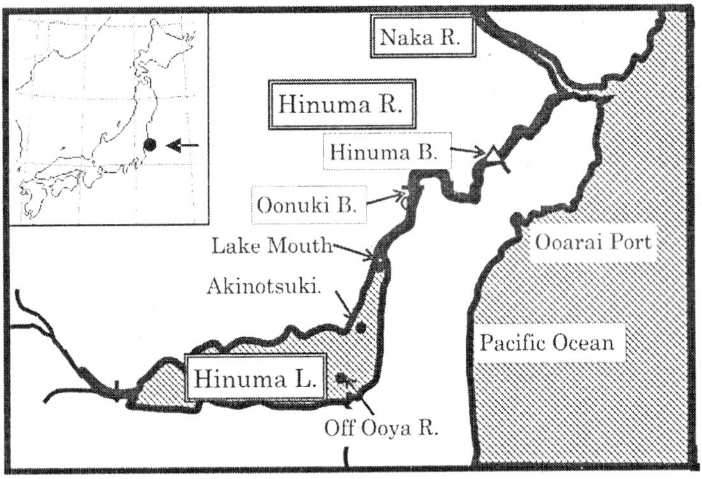

Figure 1: Hinuma river basin.

two decades. The citizens and fishermen attribute it to salinity decrease by the change of water depth in the downstream of Hinuma River, and land reclamation around the lake mouth.

3. PROCESS OF SALT WATER INTRUSION

Vertical distributions of salinity from the junction of the two rivers to the lake were measured in different tidal phases by manual salinometers from a ship. The self-registering salinometers measured temporal change of salinity at five points shown by Fig. 2 during one year, which was implemented by the cooperation between Ibaraki University, National Research Institute of Fisheries Engineering and Ibaraki Prefecture Inlandwater Fisheries Institutes. The self-registering salinometer at 11.5 km point implemented by Ibaraki Prefecture Inlandwater Fisheries Institutes has been working from August 1997.

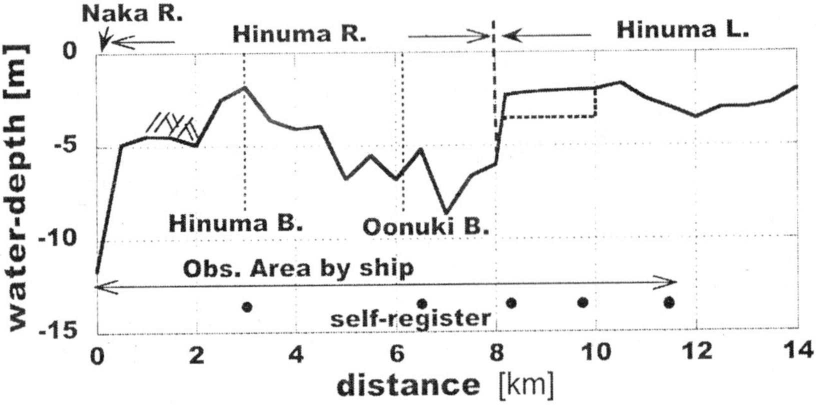

Figure 2: Longitudinal section of Hinuma river.

Strong salt water instrusions were observed on December 19, 2002 as shown in Fig. 3. Figures 3(a), (b) and (c) show the results in the end of first flood tide phase, weak ebb tide phase and next flood tide phase respectively.

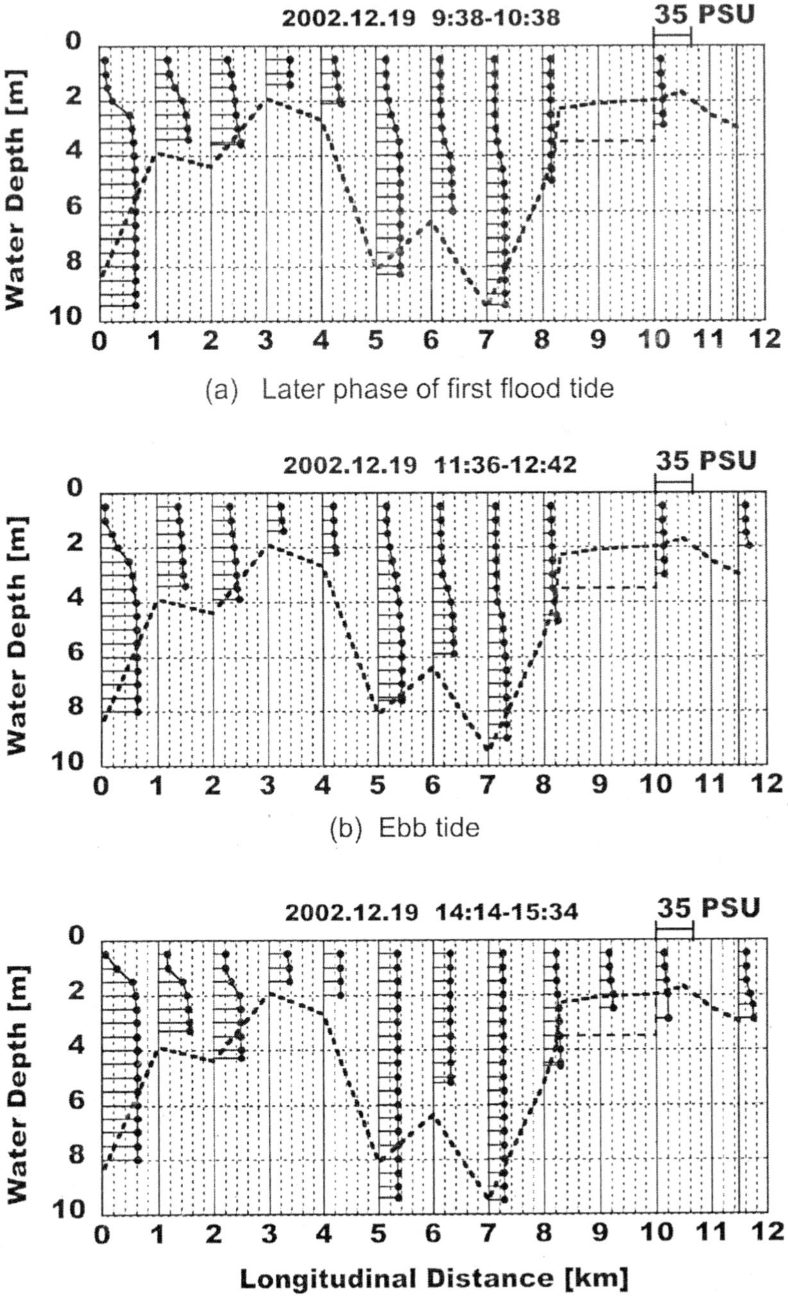

Figure 3: Salinity distribution around Hinuma river.

Salt waters start to flow into the tributary after the halocline interface in Naka River becomes higher than the bottom level of the tributary (Fig. 3(a)). This is usually delayed a few hours after the initial phase of the flood tide. The vertical profile of salinity becomes partially mixed-type at the downstream side. On the other hand, the profile changes to uniform when the salt water mass passes a quite shallow portion 3 km upstream from the junction. This salt water stalls at a deep region between 4 and 8 km and cannot further intrude into the lake in a single tide period, 12 hours. In the following ebb tide phase (Fig. 3(b)), the salt water body in this deep region has been stalled and the upper part of water body returns to downstream. The vertical profile in downstream side from 3 km becomes stronger mixed-type than that in the previous flood tide phase. In the next flood tide (Fig. 3(c)), new salt water body which is strong mixed-type comes in this deep region, and the previous salt water body stalled in this deep region intrudes to the lake. The observed results on the other day when the tidal range was large was also same as these dynamic process. When the tidal range was small, the horizontal length of the process was short and a salt water body did not intrude to the lake. A large salt water body intrudes to the centre on the Lake about ten times only in a year (Fig. 4). Figure 5 shows the temporal change of the active salinity intrusion from the lake mouth to the centre region of lake in November 2002. The observed 6.4 km point is in the tributary, the 8.2 km and 10 km point are shallow water depth around the lake mouth and the 11.7 km point is the entrance of the centre part. The periodical fluctuation of salinity following the tidal fluctuation occurred up to the shallow region, the 10 km point. When a large amount of salinity is supplied at lake mouth, the 8.2 km point, this salt water body infiltrates to the centre part of lake through the 10 km point. The remarkable phenomenon is time lag of salinity fluctuation

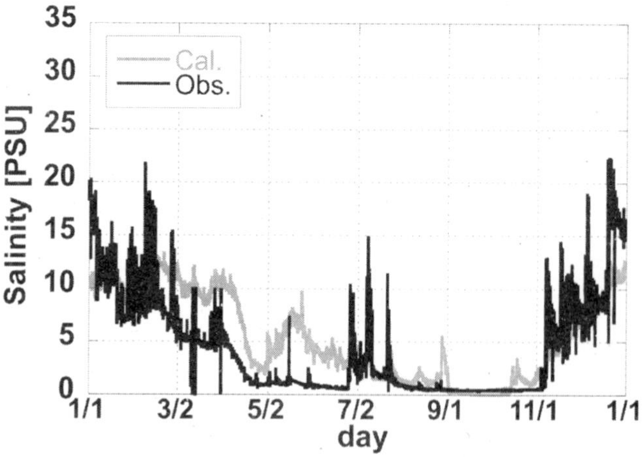

Figure 4: Salinity in Hinuma lake.
(The observed data are provided by Ibaraki Prefecture Inlandwater Fisheries Institutes)

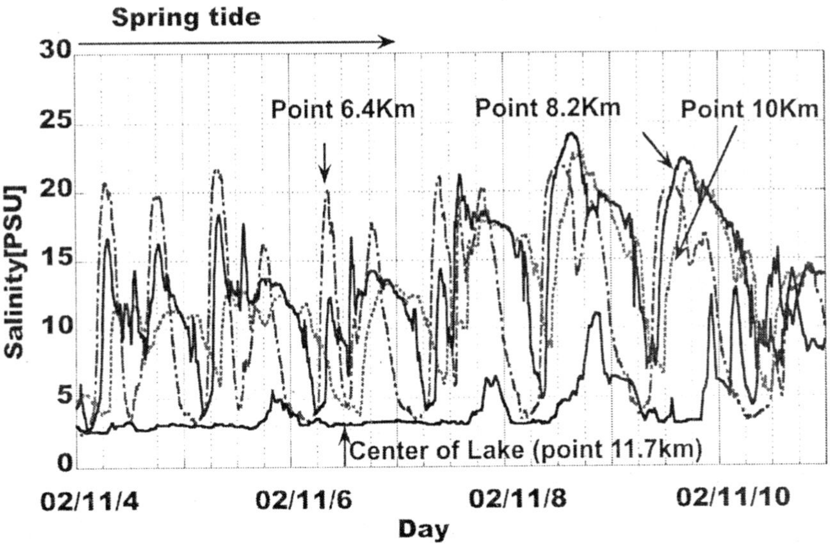

Figure 5: Temporal change of salinity into Hinuma lake.
(The observed data at Point 6.4 and 8.2 provided by National Research Institutes of Fisheries Engineering, Japan)

between the 8.2 km to 10 km point when the salt water body intrudes. The lag is a few hours. As water surface fluctuation in the lake is delayed three hours on comparing with that at the Ooarai port in Pacific Ocean, the time lag of salinity is not the phase difference of tide waves. These results indicate that the shallow water depth topography around the lake mouth act as a filter obstructing the salinity infiltration, i.e., the topography change the salinity dynamics from continuous to intermittence phenomena.

4. THREE DIMENSIONAL FLOW MODEL

The present model consists of two sub-models for flow field and density field, which are almost same as the model adapted to Tokyo bay in Japan by Mimura et al. (1998). The governing equations of flow model are continuity and momentum equations (Equations 1 and 2) and that of density model is diffusion equation of salinity (Equation 3).

$$\frac{\partial u}{\partial x}+\frac{\partial v}{\partial y}+\frac{\partial w}{\partial z}=0 \tag{1}$$

$$\frac{\partial u}{\partial t}+\frac{\partial u^2}{\partial x}+\frac{\partial uv}{\partial y}+\frac{\partial uw}{\partial z}=-\frac{g}{\rho}\frac{\partial \rho\eta}{\partial x}-fv+\frac{\partial}{\partial x}\left(A_x\frac{\partial u}{\partial x}\right)+\frac{\partial}{\partial y}\left(A_y\frac{\partial u}{\partial y}\right)+\frac{\partial}{\partial z}\left(A_z\frac{\partial u}{\partial z}\right) \tag{2}$$

$$\frac{\partial s}{\partial t}+\frac{\partial su}{\partial x}+\frac{\partial sv}{\partial y}+\frac{\partial sw}{\partial z}=\frac{\partial}{\partial x}\left(K_x\frac{\partial s}{\partial x}\right)+\frac{\partial}{\partial y}\left(K_y\frac{\partial s}{\partial y}\right)+\frac{\partial}{\partial z}\left(K_z\frac{\partial s}{\partial z}\right) \quad (3)$$

In the above equations u, v and w are velocity in x, y and z direction, h and η are water depth and water surface elevation, S is salinity, and g, ρ and f are gravity acceleration, water density and Coriolis coefficient.

The improved parts of the model compared with the previous model were that Symmetric SOR method and Donor Accepter method are employed to solve the governing equation of flow model to reduce the numerical error. Donor Accepter is a method combined with Central and up-wind difference scheme. The parameter of the combination ratio is an empirical variable to get the stable and precise solutions in each tide and river flux condition. As boundary condition, temporal records of tidal elevation at Oarai port near the river mouth in Pacific Ocean, flux rate at each river and those of wind stress were given from field observations implemented by Ibaraki prefecture and Ministry of Land, Infrastructure and Transport. No salinity measured data for this simulation exists so that the values were set as 35 PSU and 0 PSU at Ocean and river boundary, respectively.

Figure 4 also shows the capacity of salinity prediction by the model in the condition of present sea level. Calculated result is in a good agreement with observed data.

5. IMPACT OF SEA LEVEL RISE BY GLOBAL WARMING

The changes of salinity in the lake by sea-level rise were predicted by the model. Scenarios for sea-level rise were four cases as shown in Fig. 6; +9 cm (minimum), +50 cm (average) and +88 cm (maximum) following

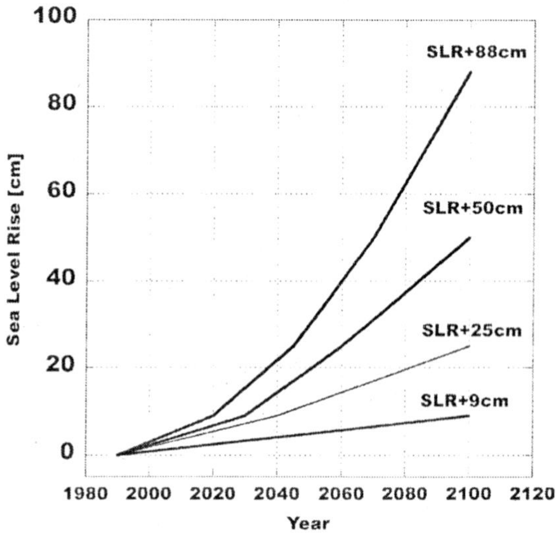

Figure 6: Four scenarios of sea level rise.

IPCC (2001) and + 25 cm which is the interpolation level. The target season was summer which is the spawning and growing season of the corbicula clam.

The calculated results of salinity at Off-Ooya River by each sea-level in 2100 are shown in Fig. 7. From the case +25 cm, the increase in salinity appears clearly. In the case +50 cm, the salinity rises much more not only along the tributary but also in the lake. The clam cannot survive under the high salinity of over 23 PSU (Nakamura, 1999), and even at present, they do not live in the ocean side from 3 km point from the junction because of this limit. Therefore, as sea-level rises, the living area of Corbicula will move upstream. In the case +88 cm, as the salinity above the bottom of the tributary almost becomes over 23 PSU, the clam will be able to live only in the lake.

The density of the average salinity below the water depth 2 m in the lake is taken for the significant salinity. This significant salinity is about 3 PSU lower than the results in Fig. 7. The temporal change of the significant salinity during sea level rising is shown in Fig. 8. The difference by the scenario of the sea level rise will appear gradually after 2020, and will become clear in about 2050. The influence of the sea level rise will not be able to be confirmed in a minimum scenario (SLR+9 cm in 2100).

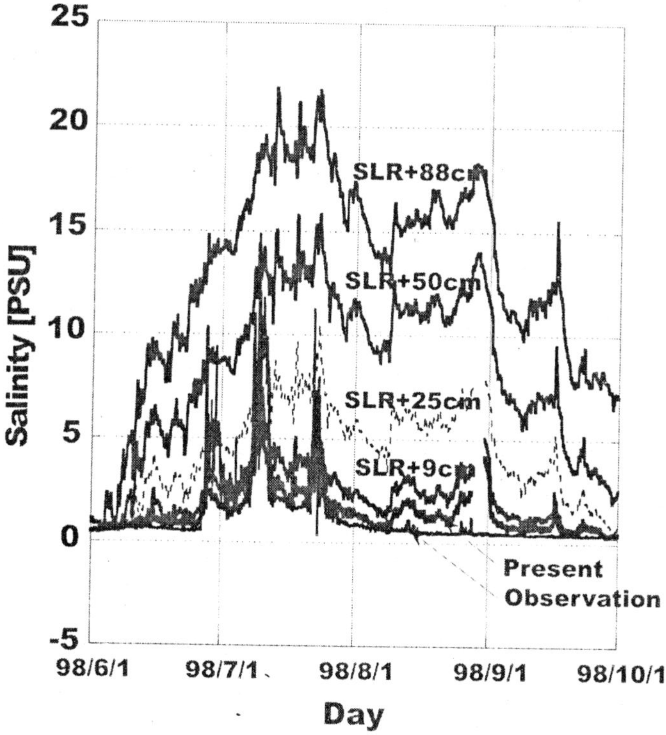

Figure 7: Impact of sea level rise.

The large change of the rate of discharge of fresh water in every year influences inflow rate of salinity very much. Table 1 shows the month-averaged salinity in the lake for five years. The difference between maximum and minimum salinity is about 9 PSU and the standard deviation of the salinity is about 4 PSU.

Table 1: Mean salinity in June at Off-Ooya River (Ibaraki Prefecture Inlandwater Fisheries Institutes)

Year	*1998*	*1999*	*2000*	*2001*	*2002*	*Average*	*Standard deviation*
Salinity [PSU]	3.29	1.77	1.83	10.99	5.32	4.643	3.83

To take account of this fluctuation, maximum or minimum salinity are estimated by following supposed equation,

$$S_{max}.S_{min} = S_{cal.}(1 \pm \frac{\sigma_{obs}}{\overline{S_{obs}}}) \tag{4}$$

in which σ_{obs} and $\overline{S_{obs}}$ are standard deviation and averaged salinity calculated from the observation data shown in Table 1, S_{cal} is calculated salinity by simulation model and S_{max} and S_{min} are predicted maximum and minimum salinity. The predicted salinity on each scenario is shown in Fig. 8. The solid, upper and lower dotted lines are the average, maximum and minimum salinity, respectively.

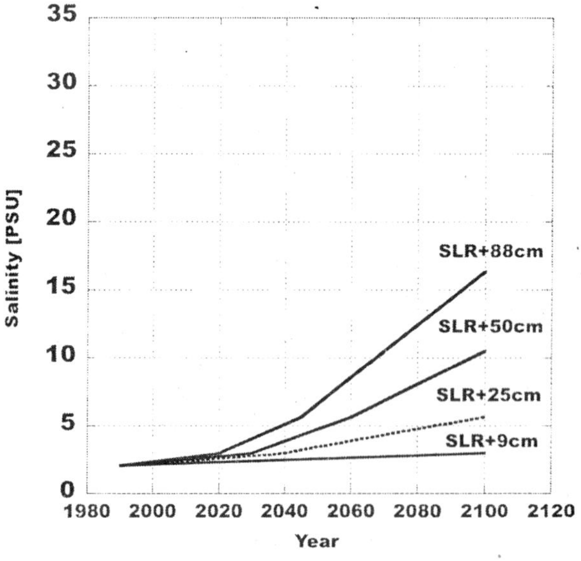

Figure 8: Temporary change of salinity in the lake by sea level rise.

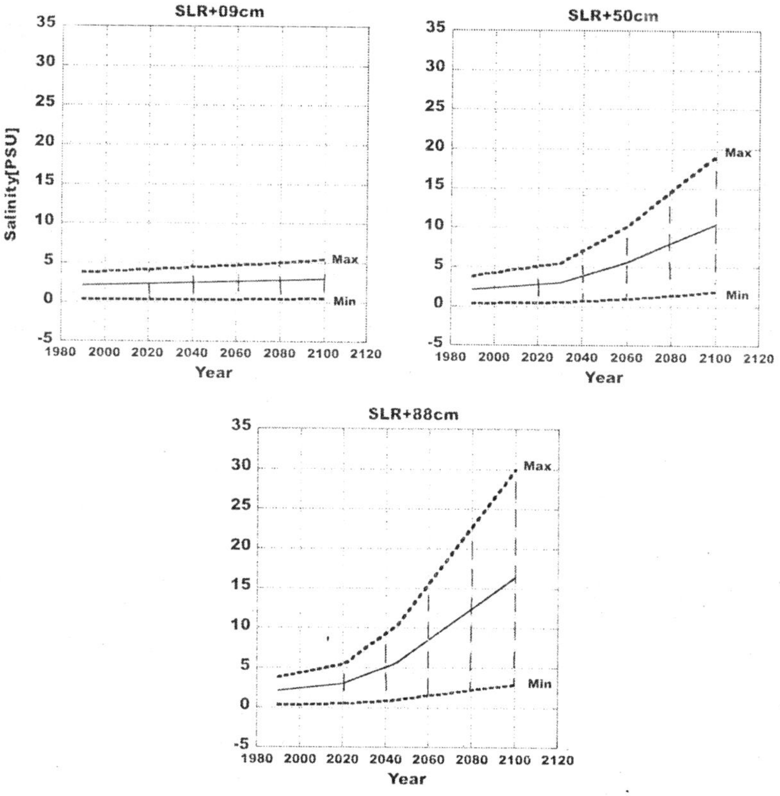

Figure 9: Temporal change of salinity in taking account of water discharge.

Though the difference between maximum and minimum salinity becomes large, we may be able to understand these results for the following reasons. When a large river flood occurred, there were no salt waters even in the river mouth because the width of the mouth in Naka River is only 200 m. On the other hand, when a water shortage occurred, salt water body arrived at the 15 km point in Naka River from the river mouth and at the 1.5 km upstream point from the lake. These results suggest us that we have to monitor the salinity rise induced by sea-level rise in taking account of these fluctuations, i.e., the effect of the flux rate of fresh water. For Hinuma basin, the evaluation term has to take 10 years which is the one cycle of the flux rate change induced by rain fall, or the evaluated salinity by sea level rise has to be calculated except the effect of the flux rate from observed salinity.

6. ADAPTATION TO SEA-LEVEL RISE

If we can raise the bottom of the river and lake in parallel to sea-level rise, salinity in this area will not change. However, as it is economically impossible to raise the bottom in all of the area, we should find the narrow sections for effective options of adaptation taking into account historical natural changes. In this study, the following four options were set;

Case A: The river bottom is raised at only the 3 km points in the tributary, where the water depth is shallowest in the river even at present. At this position, sands deposit naturally.

Case B: The artificial channel near the lake mouth is reclaimed again to put it back to the natural elevation.

Case C: The maximum water depth from the junction points of both rivers to 3 km in the tributary is set as 4 m. It may take a long time to attain this topography naturally.

Case D: Only the bottom in the tributary close to the junction rises up. At this position, the bottom elevation often changes due to flood flow in the main stream.

Through an estuary, a complete barrier is effective only for shutting out the salt water. However, this structure has been generating large destruction of ecosystem in Japan. Therefore, this option was not adapted in this study.

Figure 10 shows the temporal change of salinity at Off-Ooya River for the adaptation of Cases A and B, and Fig. 11 shows the comparison of all the adaptation capacity. Although each case reduce the salinity a little, Case-A which is the bottom up at the shallowest point in the tributary was found to be most effective among all the options to prevent salinity change induced by sea-level rise in this lake. The salinity at the upstream side of lake was reduced well. To decrease the salinity density more in the basin while keeping the above concept, we will need to take the additional adaptation in the main stream, Naka River.

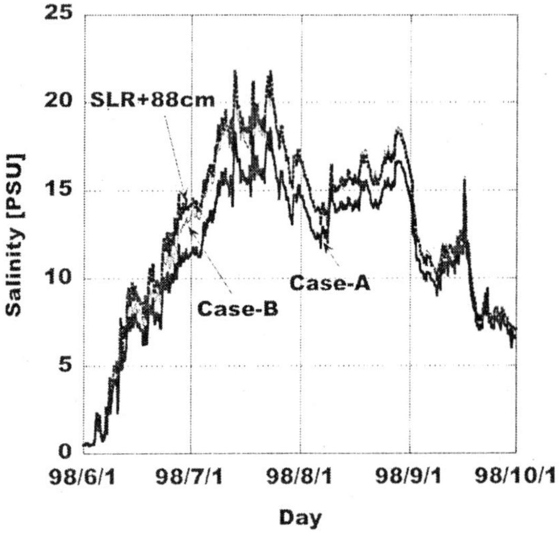

Figure 10: Effect of adaptation for salinity.

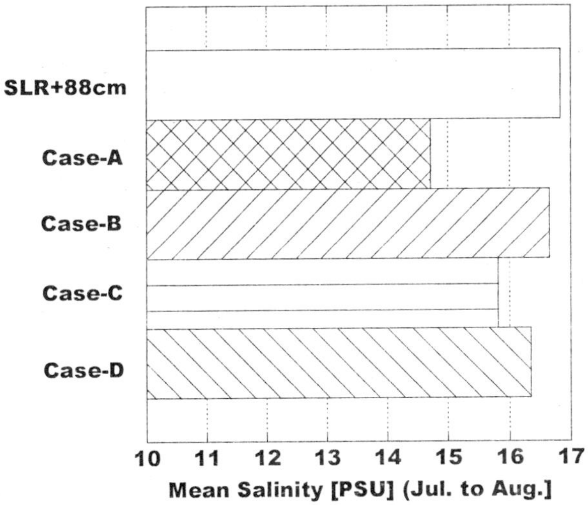

Figure 11: Comparison of salinity by adaptations.

7. CONCLUSIONS

The process of salt water body intrusion in Hinuma basin is complex due to the irregular water depth. The large salt water body is formed in the centre region of the lake only ten times in a year. The numerical simulation results, however, show that the body will intrude much according to the sea-level rise. Increasing salinity in the lake will appear after 2020 and become distinct around 2050. In the most critical scenario, which is +88 cm sea-level rise, the salinity in the tributary will become higher than the density for the clam to survive. For salt water instruction against sea-level rise, adapting the concept of historical natural change which maintains the natural process as best as possible, was not enough to reduce the salinity. Therefore, the present study suggests that we will need a support of natural power, for example natural topography change, to keep ecosystems against sea-level rise.

REFERENCES

International Panel for Climate Change, 2001. IPCC Third Assessment Report: Climate Change 2001.

Nakamura, M., 1999. Fishery of *Corbicula japonica Primes*, *Tatara publication*, 266p (Japanese).

Mimura, N., Tukada, M. and Suzuki, M., 1998. Simulation of Behavior of Oxygen-Deficit Water in Tokyo Bay by Three-Dimensional Water Quality Model. *Coastal Engineering* 1998, ASCE, pp.3575-3587.

Numerical Simulation of Salinity Structure in Chilika Lake

A. Dube, G. Jayaraman, A.D. Rao and P.K. Mohanty[1]

Centre for Atmospheric Sciences, Indian Institute of Technology Delhi
Hauz Khas, New Delhi - 110016
[1]Department of Marine Sciences
Berhampur University, Berhampur - 760 007

1. INTRODUCTION

The physical processes in Chilika Lake (Fig. 1) are complicated due to the flow regime, effluent discharge from urban runoffs and treated/untreated pollution sources. It is important to understand and analyze the individual contributory factors—changes in the landscape, climate variability, eutrophication, effects of species recruitment events etc.—since the physical, biological and chemical variability result from a combination of these effects and their interaction. The entire ecosystem of Chilika Lake is under continuous threat due to siltation, choking of the mouth connecting the lake to the sea, eutrophication, weed infestation, salinity changes and decrease in fishery resources. Hence remedial measures have to be taken in order to preserve the ecosystem of Chilika.

The present study is concerned with the seasonal circulation and salinity structures in Chilika. The high biodiversity of the lake is maintained due to this cyclic variation. It is because of the influx of fresh water during monsoon season and the inflow of sea water through the outer channel that a distinct salinity gradient exists along the lagoon. This periodic mixing of fresh and saline water helps in maintaining different species—marine, estuarine and fresh water ones—in Chilika. Also, currents regulate the supply of dissolved inorganic nutrients to the surface layer where most of the phytoplankton is found. They also control the light levels experienced by phytoplankton below the surface layer. Thus, physical exchange process plays a key role in the sustainability of the planktonic food web. The conclusions, based on a systematic study of the dynamics, salinity and ecology, should help in analyzing whether the significant improvement found in the productivity of the lake, is sustainable.

In Section 2, a brief description of the study domain is given. This is followed by Section 3 where the mathematical equations pertaining to our study domain are solved using finite difference scheme with staggered C-grid. Field observations and data required to validate the model are given in Section 4 and finally, Section 5 presents the results, based on several numerical experiments and finally in Section 6 important conclusions based on the analysis of the results of Section 5 are given. The objective is to understand the dynamics of the lake corresponding to (i) the Southwest monsoon (SWM) and Northeast monsoon (NEM) seasons and (ii) pre and post new mouth-opening conditions which will add to our initiative to improve the lake and its ecosystem.

2. DESCRIPTION OF STUDY AREA

The water spread area of the Chilika Lake varies between 1165 and 906 sq km during the monsoon and summer respectively (Siddiqi and Rao, 1995). A significant part of the fresh water to the lake comes from river Mahanadi and its distributaries (Mohanty et al., 1996). It shows an overall increase in depth of about 50 cm - 1 m due to SWM. The seasonal mode of variation in Chilika is dominant over the interannual mode (Pal and Mohanty, 2002).

Based on the physical and dynamical characteristics of the lake, the lake is divided into four compartments—northern, southern, central sectors and the outer channel (Fig. 1). The outer channel has an inlet, which was

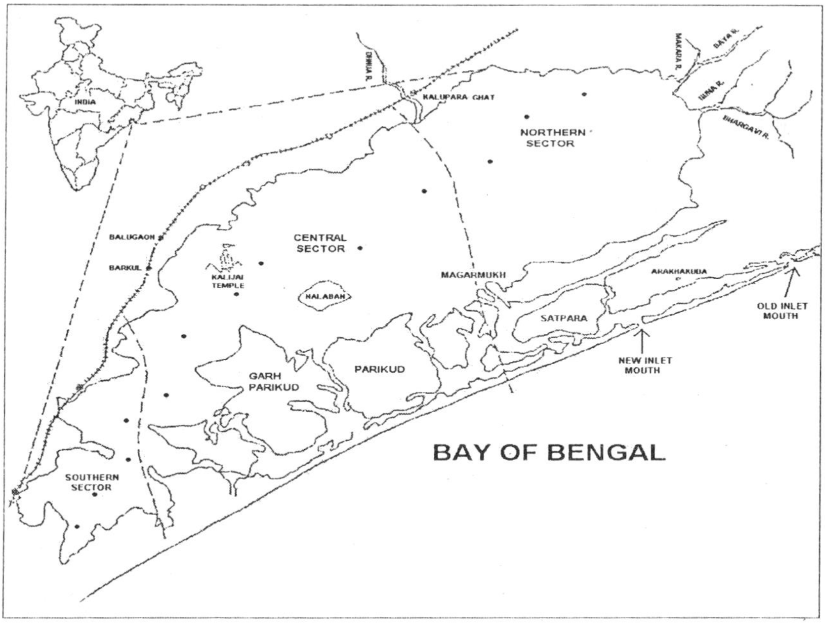

Figure 1: Map of Chilika lake showing different sectors.

the only connection of Chilika with the Bay of Bengal till September 2000 (Fig. 1). The inlet affects the exchange of water between the lake and the Bay of Bengal; it also controls the current and salinity patterns in the lake and its size determines the amount of water and salt exchanged with the Bay of Bengal.

Due to choking of the outer channel (Chandramohan and Nayak, 1994) exchange of water and salt between the lake and the sea was affected and for the preservation of Chilika ecosystem a new inlet mouth was opened near Sipakuda (Fig. 1) on 23 September 2000 (www.chilika.com). Opening of the new mouth has significantly changed the lagoon environment (Chilka, 2001) making it essential to compare the pre and post mouth-opening circulation patterns and salinity structures.

3. MATHEMATICAL FORMULATION

We use a system of rectangular Cartesian co-ordinates (Fig. 2) in which, the origin O is within the equilibrium level of the sea-surface, Ox points eastwards, Oy northwards and Oz vertically upwards. The u, v, w are the Reynolds averaged components of velocity in the directions of x, y and z respectively, f (= 5×10^{-5}) the Coriolis parameter, g (= 9.81 ms^{-2}) the acceleration due to gravity, ρ (= 1.025×10^3 kgm^{-3}) the density of the water supposed to be

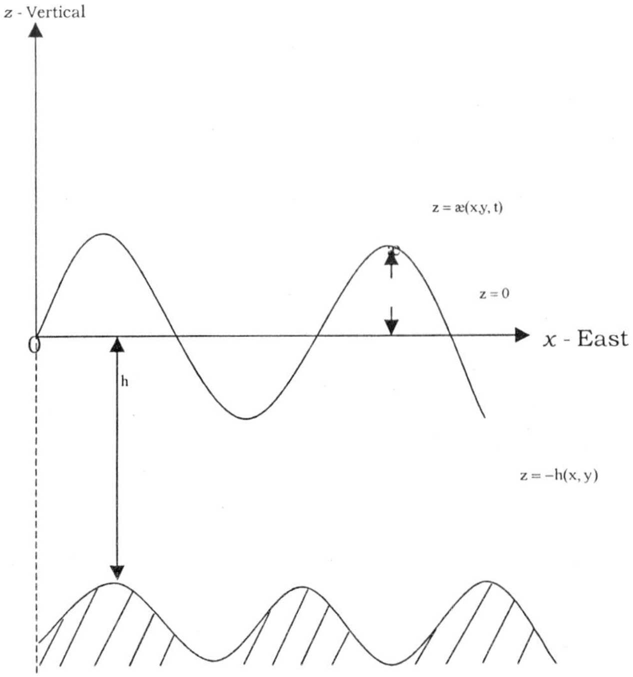

Figure 2: Coordinate representation.

homogeneous and incompressible, t time, S the salinity and K_x and K_y the kinematic eddy diffusivities in the x and y directions respectively assumed to have the same numerical value of 600 m^2s^{-1}. The (F_s, G_s) and (F_B, G_B) are the x- and y-components of the surface wind and bottom stress respectively. P_a is the surface pressure. $z = \zeta(x, y, t)$ gives the displaced position of the free surface and the position of the sea floor is given by $z = -h(x, y)$. The ratio of fractional change in density for unit change in salinity is given by $\beta(= 6.5 \times 10^{-4})$.

The lake is shallow and so we integrate vertically the basic equations of continuity and momentum to get the equations describing the dynamics of the lake as

$$\frac{\partial \zeta}{\partial t} + \frac{\partial \tilde{u}}{\partial x} + \frac{\partial \tilde{v}}{\partial y} = 0 \tag{1}$$

$$\frac{\partial \tilde{u}}{\partial t} + \frac{\partial}{\partial x}(\tilde{u}\bar{u}) + \frac{\partial}{\partial y}(\tilde{u}\bar{v}) - f\ \tilde{v} = -g\,(\zeta + h)\left(\frac{\partial \zeta}{\partial x} + \frac{1}{2}\beta\ (\zeta + h)\frac{\partial \bar{S}}{\partial x}\right) + \frac{1}{\rho}\left[F_S - F_B\right] \tag{2}$$

$$\frac{\partial \tilde{v}}{\partial t} + \frac{\partial}{\partial x}(\tilde{v}\bar{u}) + \frac{\partial}{\partial y}(\tilde{v}\bar{v}) + f\ \tilde{u} = -g\ (\zeta + h)\left(\frac{\partial \zeta}{\partial y} + \frac{1}{2}\beta\ (\zeta + h)\frac{\partial \bar{S}}{\partial y}\right) + \frac{1}{\rho}\left[G_S - G_B\right] \tag{3}$$

$$\frac{\partial\left((\zeta + h)\bar{S}\right)}{\partial t} + \frac{\partial\left(\bar{u}(\zeta + h)\bar{S}\right)}{\partial x} + \frac{\partial\left(\bar{v}(\zeta + h)\bar{S}\right)}{\partial y} = \frac{\partial}{\partial x}\left[K_x\,(\zeta + h)\frac{\partial \bar{S}}{\partial x}\right] + \frac{\partial}{\partial y}\left[K_y\,(\zeta + h)\frac{\partial \bar{S}}{\partial y}\right] \tag{4}$$

where $\tilde{u} = (\zeta + h)\bar{u}$ and $\tilde{v} = (\zeta + h)\bar{v}$ are new prognostic variables, $(\zeta + h)$ gives the total depth of the water column and over-bars denote depth-averaged values given by

$$(\bar{u}, \bar{v}) = \frac{1}{(\zeta + h)}\int_{-h}^{\zeta} (u, v)dz \text{ and } (\bar{S}) = \frac{1}{(\zeta + h)}\int_{-h}^{\zeta} (S)dz \tag{5}$$

A parameterization of the bottom stress is made by the conventional quadratic law.

$$\left.\begin{aligned} F_B &= \rho\ c_f\ \bar{u}\ (\bar{u}^2 + \bar{v}^2)^{1/2} \\ G_B &= \rho\ c_f\ \bar{v}\ (\bar{u}^2 + \bar{v}^2)^{1/2} \end{aligned}\right\} \tag{6}$$

where $c_f = 1.25 \times 10^{-3}$ is an empirical bottom friction coefficient.

Surface shear stress due to wind is usually computed using a bulk aerodynamic formula

$$\left.\begin{aligned} F_S &= C_D\ \rho_a\ u_a\ \left(u_a^2 + v_a^2\right)^{1/2} \\ G_S &= C_D\ \rho_a\ v_a \left(u_a^2 + v_a^2\right)^{1/2} \end{aligned}\right\} \tag{7}$$

where C_D ($=1.125 \times 10^{-3}$) is the drag coefficient, ρ_a ($=1.176$ kgm^{-3}) is the density of air and (u_a, v_a) are the wind velocity components in x and y directions, respectively.

The analysis area covers an extent of about 60 km (east-west) and 50 km (north-south) and in order to incorporate all the islands, the coastal boundaries and the open-channels, at least a minimum grid interval of about 750 m is required. Accordingly, the real boundary of Chilika Lake is approximated as closely as possible by a stair-step boundary (Fig. 3). With this grid specification, it was found that computational stability could be maintained with a time step of 60 sec. The coastal boundary in the model is taken to be a vertical sidewall through which there is no flux of water except for the two open channels connecting the lake to the sea.

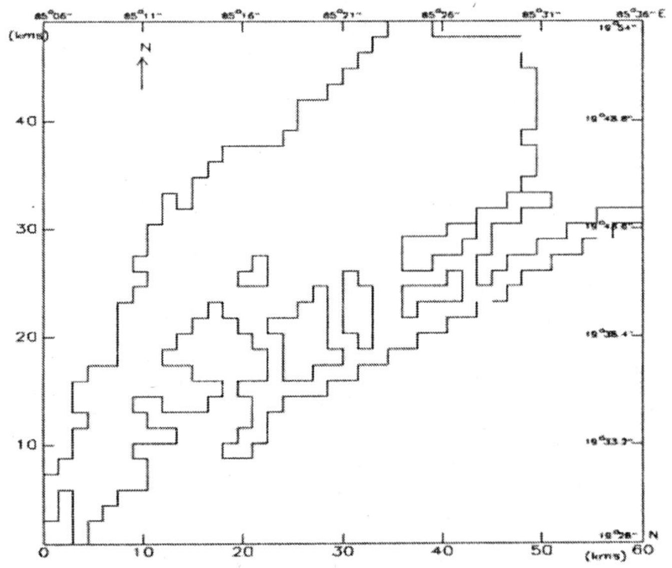

Figure 3: Representation of the boundary by orthogonal stair steps.

3.1 Boundary and Initial Conditions

Theoretically the only boundary condition needed in the vertically integrated system is that the normal transport vanish at the coast, i.e.,

$$u \cos \alpha + v \sin \alpha = 0 \quad \text{for all } t \geq 0 \tag{8}$$

where α denotes the inclination of the outward directed normal to the x-axis. It then follows that $u = 0$ along the y-directed boundaries and $v = 0$ along the x-directed boundaries. In addition to the boundary conditions for the hydrodynamic model, appropriate boundary conditions have to be specified for salinity equation. Along with the normal transport, the diffusive flux of salinity at the lateral boundaries must vanish i.e.,

$$\frac{\partial S}{\partial n} = 0 \text{ where } n \text{ is the unit normal to the lateral boundary.} \tag{9}$$

At the tidal inlets, a sinusoidal tide is prescribed and given by

$$\zeta(t) = a \cos\left(\frac{2\pi t}{T}\right) \tag{10}$$

is used. The a is the tidal amplitude and T is the time period of M_2 tide which is equal to 12.4 hrs approximately. The two tidal inlets of the lake are very nearby (17 km apart) and therefore, the tides at both the tidal openings are considered to be in phase and having the same amplitude a.

The fresh water flux at the river opening is provided in terms of the velocity of fresh water entering into the Chilika basin from the rivers. This is calculated according to the following formula

$$u(x, y) = \frac{q}{h(x, y)\, L} \tag{11}$$

where q (3300 million cum) is the amount of fresh water in cubic metres entering into the Chilika basin obtained by observations made during the SWM and L is the length of the inlet.

For the solution of the equations (1-4) subject to the boundary conditions a finite difference scheme with staggered C-grid is used (Mesinger and Arakawa, 1976).

4. OBSERVATIONS

Observations on hydrographic parameters correspond to SWM and NEM covering about 33 to 37 stations in the body of the lake after the opening of the new mouth. Observations before the opening of the new inlet were obtained from Mohanty et al. (2001). Methodology for observation and analysis of most of the parameters was adopted from Parsons et al. (1984). The observations are given in Tables 1(a, b). All the hydrographic parameters show a wide spatial and temporal variability. There is almost a uniform

Table 1a: Summary of hydrographic parameters in Chilika lake (Values within bracket indicate the ranges) Southwest Monsoon (pre mouth-opening)

	Northern	*Central*	*Southern*	*Channel*
Depth (m)	0.95	2.2	2.4	1.6
	(0.3-1.6)	*(1.6-2.8)*	*(1.5-3.3)*	*(1-4)*
Salinity (ppt)	0.85	6.9	9.6	*7.63*
	(0.4-1.3)	*(4.6-9.2)*	*(8.0-11.2)*	
Southwest Monsoon (post mouth-opening)				
Depth (m)	1.7	2.0	2.5	2.7
	(1.1-3.4)	*(1.0-3.1)*	*(1.0-3.6)*	*(1.0-4.3)*
Surface Water Current (cm/s)	65.2	66.9	57	110.7
	(33-117)	*(42-88)*	*(0-98)*	*(88-149)*
Salinity (ppt)	0.9	5.6	18.0	*3.0*
	(0.4-4.4)	*(0.4-12.7)*	*(12.7-20.2)*	*(1.8-3.6)*

Table 1b: Summary of hydrographic parameters in Chilika Lake (Values within bracket indicate the ranges and * indicates the unavailability of data)

Northeast Monsoon (pre mouth-opening)				
	Northern	*Central*	*Southern*	*Channel*
Depth (m)	0.8	1.7	2.1	1.5
	(0.2-1.4)	*(0.7-2.6)*	*(1.2-3.0)*	*
Salinity (ppt)	3.75	8.46	8.95	9.9
	(0.65-6.85)	*(7.84-9.08)*	*(8.39-9.42)*	*
Northeast Monsoon (post mouth-opening)				
Depth (m)	0.6	1.7	2.2	2.3
	(0.6-0.7)	*(0.7-3.0)*	*(1.9-2.4)*	*(1.1-4.1)*
Surface Water Current (cm/s)	3.7	1.3	4.3	22.3
	(0.00-25.0)	*(0.00-5.4)*	*(2.0-7.7)*	*(18.1-26.1)*
Salinity (ppt)	5.2	5.7	7.4	14.6
	(4.9-5.5)	*(4.4-7.7)*	*(6.8-7.8)*	*(13.4-16.3)*

decrease in the depth by about 1 m during the NEM as compared to SWM. The model bathymetry (Fig. 4) is generated based on observations from January 2002. Magnitude of current velocity is observed to be much less during NEM as compared to SWM. Maximum velocity zone is the channel (tidal effect) during NEM and for SWM the fresh water inlet.

Salinity variation in the lagoon is significant which ranges from 0.2 ppt to 22 ppt in the pre mouth-opening condition and from 0.4 to 33 ppt in the post mouth-opening condition during summer. Fresh water influx causes a reduction in salinity during SWM.

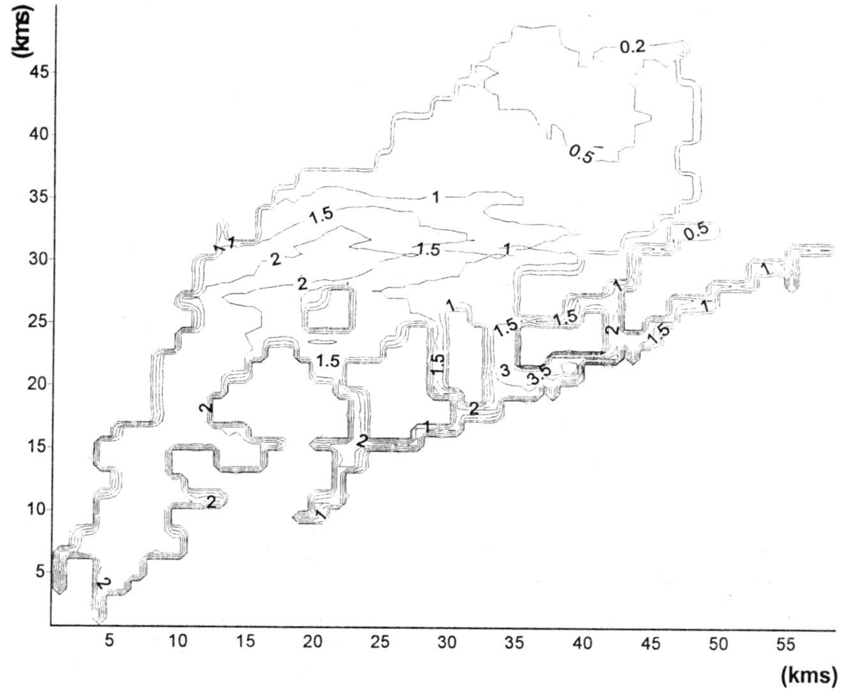

Figure 4: Bottom topography (m) Chilika lake.

5. RESULTS AND DISCUSSIONS

The 2D depth averaged model is used to obtain information on the response of Chilika waters to (i) wind forcing and (ii) tidal effects. The model is forced by the climatological mean winds representative of the months of July and January (representative months for SWM and NEM respectively). During January, the direction of the wind is almost North East (40°) with uniform speed of 2 m/s and for July the direction is almost South West (200°) with a uniform speed of 6.5 m/s over the entire lake area (Hastenrath and Lamb, 1979). At the open sea boundary, a sinusoidal tide of amplitude a (1.12 m) and salinity of 33 ppt is considered (Nayak et al., 1998). Fresh water influx from the rivers is also included in the model for SWM. Computational steady state was achieved in 25 tidal cycles and the results discussed below were obtained after steady state solution was reached. Our earlier study (Jayaraman et al. 2005) deals at length with the circulation in the lake and we shall discuss here the resulting salinity patterns.

The results are discussed separately for July and January corresponding to (i) pre mouth-opening and (ii) post mouth-opening conditions. Comparisons are later made and conclusions drawn in the next section for (i) the two different seasons and (ii) pre and post mouth-opening conditions.

5.1 SOUTHWEST MONSOON (SWM)

5.1.1 Old Tidal Inlet

Circulation and salinity in Chilika are affected by three main factors during the SWM namely wind, tide and fresh water influx. The results obtained from the numerical experiments performed with the above forcing factors are analyzed for mean circulation and salinity where the mean was taken over a tidal period after the steady state was achieved.

Figure 4 shows the residual circulation, which is ebb; hence proving that ebb current is stronger than flood—a result confirmed by observations (the peak flood current velocity is 45 cm/s and the peak ebb current velocity is 55 cm/s). The currents are almost uniform (1 to 5 cm/s) in the main body of the lake except for the northern sector due to fresh water influx (10 cm/s to 45 cm/s), the channel due to tidal influence (10 to 20 cm/s) and southern sector which has very low current velocity of about 1 cm/s.

Figure 5 shows that the region of maximum salinity is the channel and there is a steady decrease in salinity from a maximum of 33 ppt near the

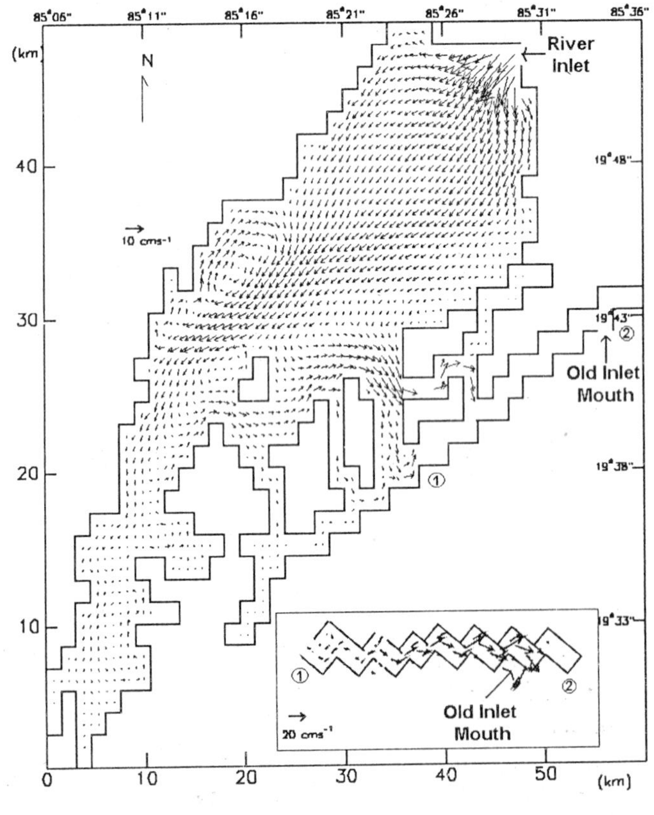

Figure 5: Mean circulation as a result of July wind, tide and freshwater influx.

inlet to 10 ppt near the central sector. Salinity progresses in central sector in the form of a plume where it ranges from 8 to 6 ppt. Further into the central sector the salinity distribution is almost uniform (6 to 4 ppt). The northern and southern sectors show low salinity values ranging from 1 to 4 ppt. The salinity ranges in different sectors match well with the observed ranges in the corresponding sectors.

5.1.2 New Tidal Inlet

In this experiment, the new tidal inlet is also included in the model. The peak flood and ebb tide velocities are 95 cm/s and 110 cm/s respectively and are observed near the new inlet mouth. The circulation profile is almost the same as with one inlet showing that the residual circulation in the lake is not affected qualitatively due to the opening of the new mouth.

Figure 6 shows the mean salinity distribution in Chilika. Salinity values show an overall increase due to the new inlet mouth. The outer channel shows a salinity range of 33 to 10 ppt. A large salinity plume dominates the central sector and has salinity range of 10 to 8 ppt. The effect of fresh water influx is clearly seen in the northern sector where the salinity is almost zero at the inlet but increases steadily as we move away from the fresh water

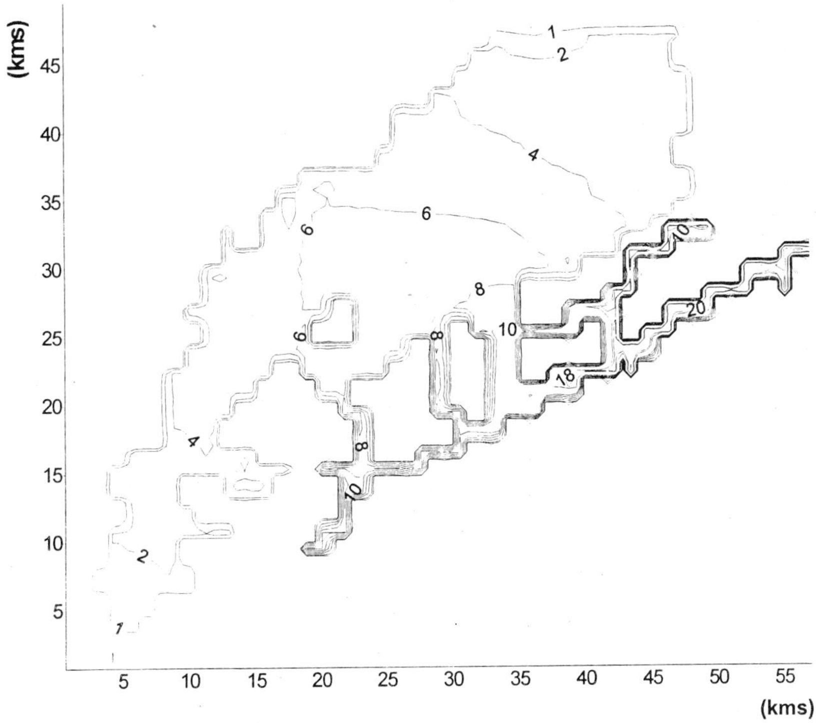

Figure 6: Horizontal salinity in ppt as a result of July wind, tide and freshwater influx (old inlet).

inlet. Northern sector shows higher salinity values (8 to 2 ppt) from the previous case of one mouth. It can be observed that the new tidal inlet affects salinity values throughout the main body of the lake except in the northernmost and the southernmost regions.

5.2 Northeast Monsoon (NEM)

5.2.1 Old Tidal Inlet

Forcing factors affecting circulation and salinity in Chilika during NEM are wind and tides. Three factors which differentiate NEM from SWM are (i) no appreciable fresh water influx, (ii) lesser depth and (iii) change in wind magnitude and direction.

Figure 8 shows the mean circulation for NEM. The mean profile is ebb which implies that ebb current is stronger than the flood (peak flood and ebb current velocities are 45 cm/s and 50 cm/s respectively). In almost the entire main body of the lake, velocity is uniform (1 cm/s). In the central sector it

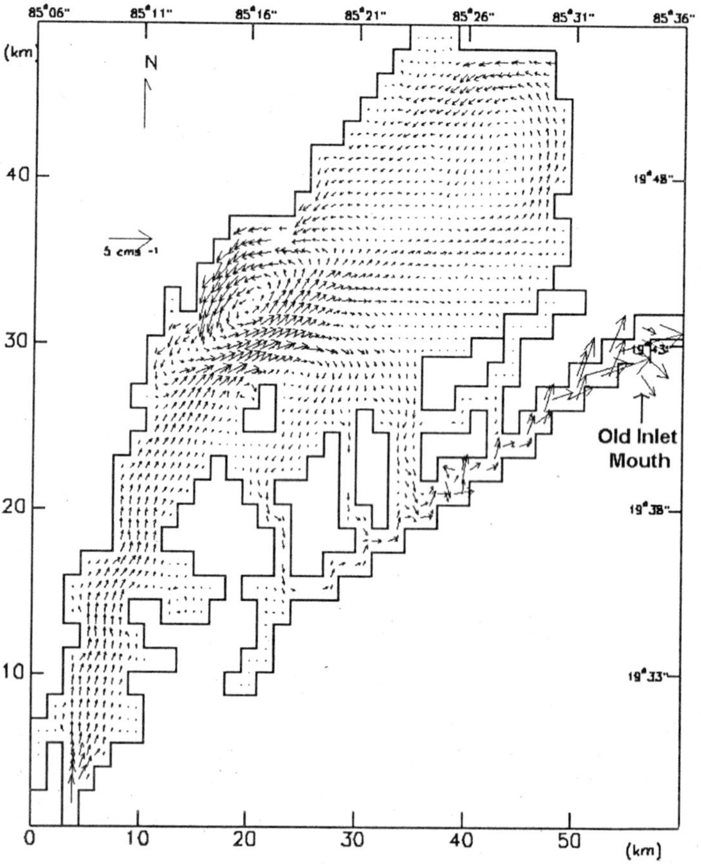

Figure 8: Mean circulation as a result of January wind and tide (old inlet).

goes slightly higher and takes a value of 2 cm/s. Channel is seen to be the region of maximum velocity where it ranges from 1 to 10 cm/s.

Figure 9 depicts the salinity structure in Chilika during NEM. Maximum salinity is observed in the channel (33 to 14 ppt). Salinity values show a steady decrease as we proceed towards the central sector where it takes on a plume structure and has a range from 12 to 10 ppt. Central sector has a salinity range of 12 ppt near the channel to 6 ppt near the northern and southern sectors. Northern sector has a salinity range of 6 ppt (near the central sector) to 1 ppt whereas the southern sector is the lowest salinity zone (4 to 1 ppt). Salinity values show an increase from their values during SWM due to a lack of fresh water influx during NEM.

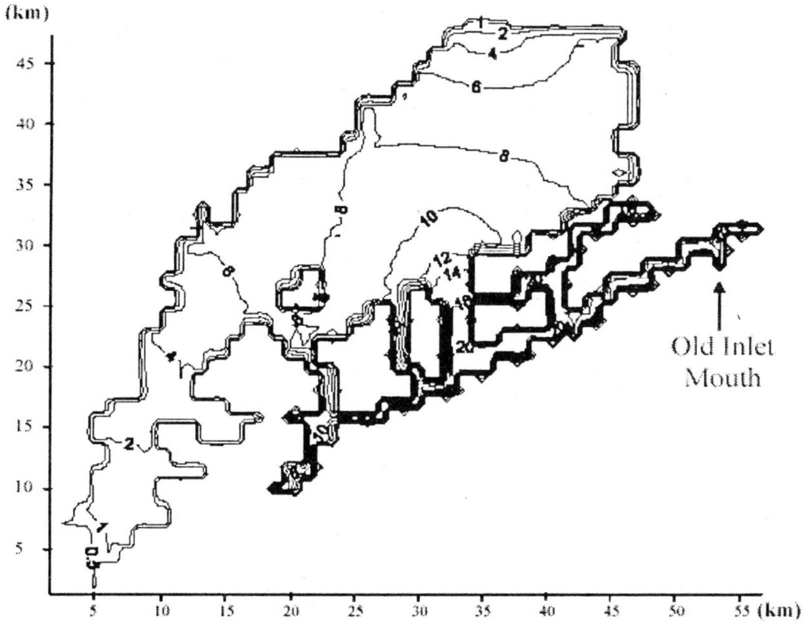

Figure 9: Horizontal salinity in ppt as a result of January wind and tide (old inlet).

5.2.2 New Tidal Inlet

Ebb current is seen to be stronger than the flood current from the mean circulation profile which is ebb (peak flood and ebb current velocities are 81 cm/s and 95 cm/s respectively). The channel shows a velocity range of 5 to 10 cm/s. Velocity in the main body of the lake is the same as the pre mouth-opening case which shows that the residual circulation in the main body of the lake remains unaffected by the opening of the new inlet mouth.

Figure 10, the mean salinity structure, shows that maximum salinity is observed in the channel (33 to 30 ppt). At Magarmukh salinity decreases to 26 ppt (as against 16 ppt observed with one inlet). The progress of salinity

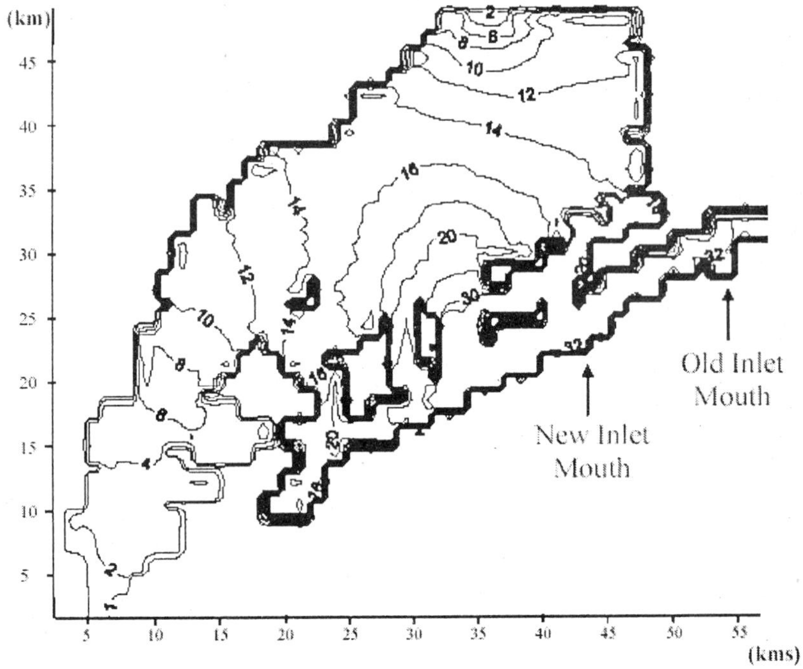

Figure 10: Horizontal salinity in ppt as a result of January wind and tide (both inlets).

into the central sector is in the form of a plume which shows a rapid variation of salinity with values ranging from 26 to 16 ppt. Northern sector shows an increased salinity range (14 to 2 ppt) which is also due to a lack of fresh water influx. Among all the other sectors the southern sector shows the lowest salinity (10 to 2 ppt). In general the way salinity progresses from the channel into the main body of the lake remains the same as from the case of one mouth except that the salinity values show an overall increase throughout the lake area due to the increased tidal influx.

6. CONCLUSIONS

A two-dimensional numerical model was developed to simulate circulation and salinity structure in Chilika Lake. Numerical experiments show that bottom topography besides wind and tides plays an important role in determining circulation patterns in the Lake. Firstly ebb current always remains stronger than flood irrespective of the seasons. The channel is the deepest region of the lake (1 to 4 m) whereas the central sector is shallower as compared to the channel (1 to 2 m). This gradient facilitates the ebb in Chilika.

Secondly two big eddies are formed in the central and the northern sectors due to the depth gradient in these regions. These two results validated by observations are confirmed to be caused by the depth gradient since the numerical experiment with uniform depth failed to show these results.

It is also seen by the numerical experiments that wind forcing is effective only in the main body of the lake whereas tidal forcing is effective only in the channel. Though the second opening has helped in increasing the tidal influx and hence the salinity, its influence is not felt in the interior of the lake due to the constriction of flow area between the lake and the channel area near Magarmukh. The simulated results have been validated against the limited observations and found qualitatively in good agreement.

Seasonal studies based on our model can help in understanding whether the observed significant improvement in the biological productivity of the lake after post mouth opening is sustainable.

ACKNOWLEDGEMENTS

Part of the work reported in this paper was carried out under a sponsored project funded by the Department of Ocean Development, Government of India.

REFERENCES

Chandramohan, P. and Nayak. B.U., 1994. A study for the improvement of the Chilka lake tidal inlet, East Coasts of India. *Journal of Coastal Research*, **10:** 909-918.

Chilka, 2001. Chilka - A New Lease of Life. Chilika Development Authority, Bhubaneshwar, Orissa, 13P.

Hastenrath, S. and Lamb, P.J., 1979. Climatic Atlas of the Indian Ocean, Part-1: Surface Climate and Atmospheric Circulation. The University of Wisconsin Press, viii-xi.

Jayaraman, G., Rao, A.D., Dube, A. and Mohanty, P.K., 2006. Numerical Simulation of Circulation and Salinity Structure in Chilika Lagoon. In Press *Journal of Coastal Research.*

Mesinger and Arakawa, 1976. Numerical Methods Used in Atmospheric Models. GARP Publication. Series, 17, **1:** 1-64.

Mohanty, P.K., Dash, S.K., Mishra, P.K. and Murty, A.S.N., 1996. Heat and Momentum Fluxes Over Chilika: A Tropical Lagoon. *Indian Journal of Marine Sciences*, **25:** 184-188.

Mohanty, P.K., Pal, S.R. and Mishra, P.K., 2001. Monitoring ecological conditions of a coastal lagoon using IRS data: A case study in Chilika, east coast of India. *Journal of Coastal Research*, **34:** 459-469.

Nayak, B.U., Ghosh, L.K., Roy, S.K. and Kankara, S., 1998. A study on hydrodynamics and Salinity in the Chilika Lagoon. Proceedings of Chilika Development Authority, 31-47.

Pal, S.R. and Mohanty, P.K., 2002. Use of IRS-1B data for change detection in water quality and vegetation of Chilika lagoon, east coast of India. *International Journal of Remote Sensing*, **23(6):** 1027-1042.

Parson, T.R., Maita, Y. and Lalli, C.M., 1984. A manual of chemical and biological methods for seawater Analysis. Pergamon press, New York, 173 P.
Siddiqi, S.Z. and Rama Rao, K.V., 1995. Limnology of Chilka Lake. *In:* Fauna of Chilka

Lake: Wetland Ecosystem, Series 1, Edited by the Director, Zoological Survey of India (Calcutta), 11-136.
www.chilika.com: Official Homepage of the Chilika Development Authority.

The Role of Benthos and Epiphyte on the Material Cycle in Akkeshi Lake, Japan

Michio J. Kishi and Yuko Oshima[1]

Faculty of Fisheries Sciences, Hokkaido University
N13W8, Sapporo, Hokkaido 060-0813, Japan
[1]PASCO Co. Ltd., 2-32-1 Yoga, Setagaya-ku
Tokyo 158-0097, Japan

1. INTRODUCTION

Akkeshi Lake is located in Hokkaido, the northern island of Japan. The lake is connected to Akkeshi Bay, which is a part of North Pacific (Figs 1 and 2). The average depth of the lake is 2 m. Oshima et al. (1999) formulated an ecological-physical coupled model and has explained the role of eel grass

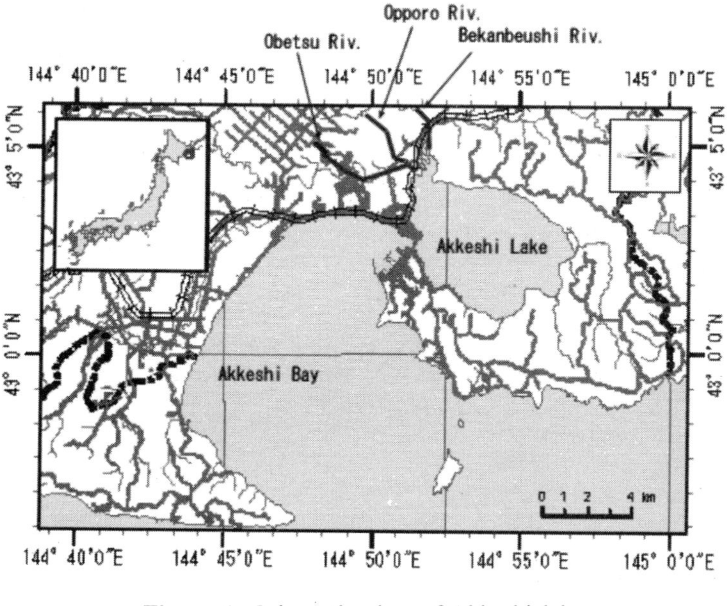

Figure 1: Schematic view of Akkeshi lake.

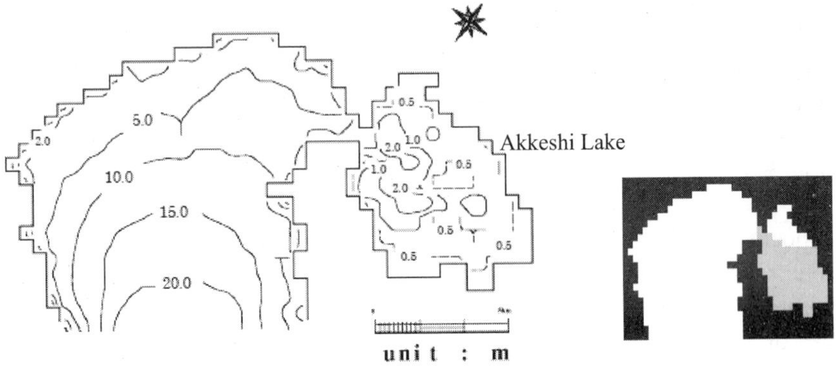

Figure 2: Bottom topography of the model area. Grey area corresponds to the box, where the budget of nutrient is calculated.

in the marine ecosystem. The major current that influence the lake is mainly caused by M_2 tide and also wind driven circulation (Oshima et al., 1999). There are three main rivers that are connected to the lake and their average nutrient load values are given in Tables 1 and 2. The oyster and clam cultures are carried out in the lake, and it has been earlier observed that epiphytes as well as cultured clam play important role in the material flow of the lake. The purpose of the present work is to reveal the role of epiphyte and clam in the lake ecosystem using three-dimensional ecological physical coupled model.

Table 1: Notation used in the physical model and their values

Horizontal gird size	f C F C	500 m
Time interval	f C	20 sec.
Average amplitude of tide		0.5 m
Period of tide		12 hour
Vertical gird size		0.5.0.5.2.0.2.0.5.0.5.0.5.0. bottom m

Table 2: Open boundary condition of ecosystem model

Layer	CHL	NH_4	NO_3	PO_4	PON	DON	POP	DOP	ZOO
1	4.77	4.32	0.28	0.45	0.40	3.00	0.04	0.30	0.40
2	4.87	4.02	0.30	0.45	0.40	3.00	0.04	0.30	0.40
3	5.12	3.28	0.34	0.46	0.40	3.00	0.04	0.30	0.40
4	5.51	2.09	0.41	0.47	0.40	3.00	0.04	0.30	0.40
5	6.76	1.78	0.71	0.71	0.40	3.00	0.04	0.30	0.40
6	7.94	1.37	1.68	0.90	0.40	3.00	0.04	0.30	0.40
7	8.07	0.54	2.39	0.87	0.40	3.00	0.04	0.30	0.40
8	8.07	0.54	2.39	0.87	0.40	3.00	0.04	0.30	0.40
Unit	µg/l	µM	µM	µM	µM	µM	µM	µM	µM

2. MODEL DESCRIPTION

The present ecological and physical models (3-D model) are based on Oshima et al. (1999). The epiphyte and clam, which have major roles in the material cycles in Akkeshi Lake (Winata, 2001) are added to that ecosystem model (Fig. 3). Iizumi et al. (1995) has pointed out that the main source of nitrate and silicate is from the river. The time dependent feature of nitrate concentration of rivers falling in Akkeshi Lake is shown in Fig. 4. Obetsu river, that flows along the urban area, is the major source of nitrate and records high values just after a heavy rainfall and might be caused by flushing out of surface dust of town. Bekanbeushi river, which mainly flows along the natural forest, indicates low nitrate concentration just after the rainfall and might be caused by dilution of nutrient. During simulation, input of nutrients from river and inflow of rivers are taken into consideration based on the observations. The number of clam is 3200 individuals m^{-2} and biomass of eelgrass is 200 g m^{-2} in April and increases linearly to 400 g m^{-2} in June. Epiphyte attached to eelgrass is taken into consideration, and their distribution coincides with that of eelgrass. Biomass of epiphyte has been observed by a student of Hokkaido University (unpublished) and is supposed to be 20 mg

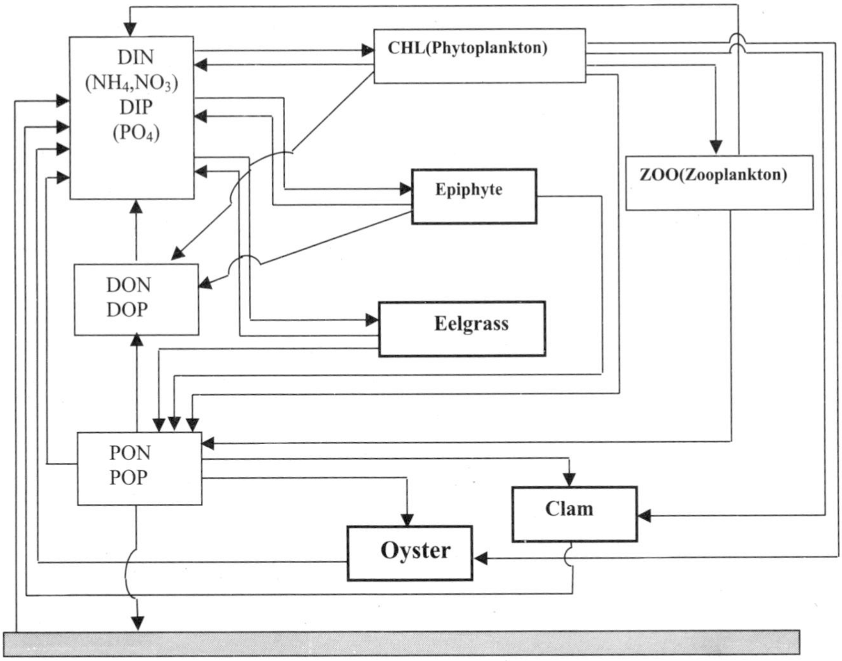

Figure 3: Schematic view of ecosystem model (DIN: dissolved inorganic nitrogen, DIP: dissolved inorganic phosphorous, DON: dissolved organic nitrogen, DOP: dissolved organic phosphorous, PON: particulate organic nitrogen, POP: particulate organic phosphorous).

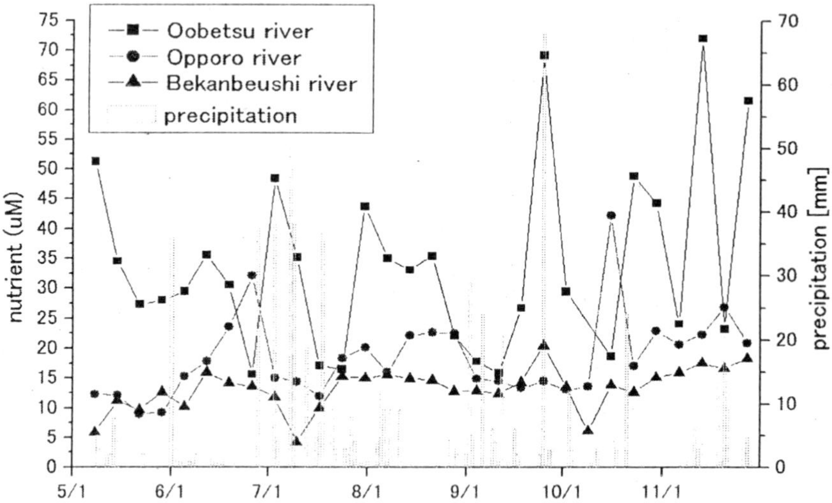

Figure 4: Nitrate concentration of rivers which flow into Akkeshi Lake and precipitation near Akkeshi in 2003.

Chl-a m^{-2} in April (Hasegawa, personal communication). Their biological parameters are assumed to be same as phytoplankton (Iizumi, personal communication). The model was run under the average meteorological condition in April for 100 days in order to obtain quasi-steady state. After that, observed wind force, river discharge and solar radiation were the inputs from May to October 2003. The analysis was made for the results averaged over above six months.

3. RESULTS AND DISCUSSION

A time dependent feature in each ecological compartment in the centre of the lake during spin up is observed (Fig. 5). This clearly shows that after 30 days, all compartments seem to reach the steady states. Thus, the results can be used after 100 days spin up as the initial condition of simulation. Figures 6 and 7 show the daily material cycles in the average simulation period condition. The inflow to clam is much larger than that to oyster because the biomass of clam is much larger than that of oyster. It is also clearly observed that clam grazes not only phytoplankton but also Particulate Organic Nitrogen (PON), the source of which is epiphyte and eelgrass. As shown in these figures, epiphyte uptakes nutrients twice as large as phytoplankton does. Epiphyte plays an important role in the ecosystem of the lake.

The horizontal distributions of ammonium and chlorophyll-a (converted from nitrogen to chlorophyll, using a C:N ratio of 133:17 and a C:Chlorophyll-a ratio of 50:1) at the sea surface as well as observed value in April, 2003 (Mukai et al., unpublished) are shown in Figs 8 and 9. As shown in Fig. 9,

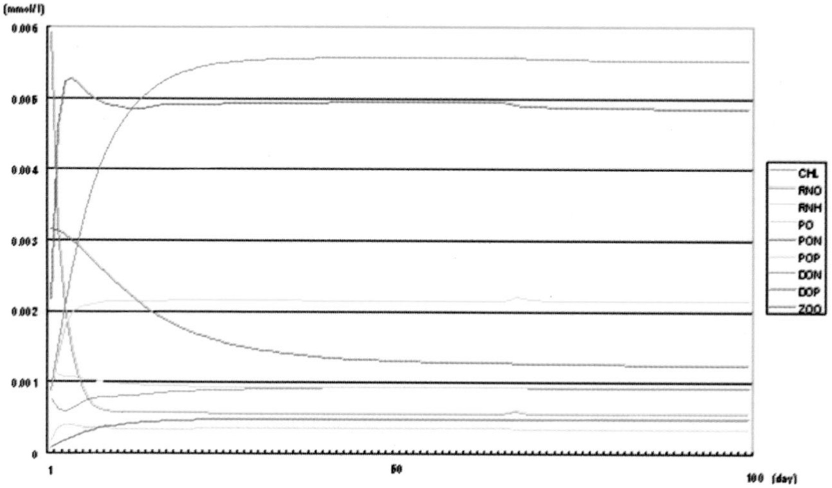

Figure 5: Time dependent value of each compartment at the central lake.

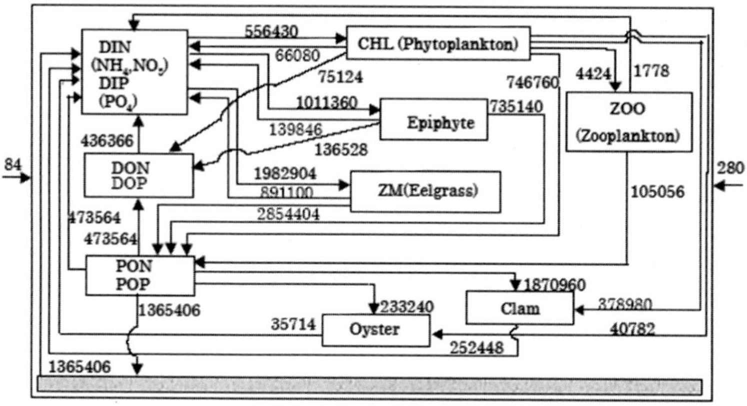

Figure 6: Daily averaged nitrogen flow among compartments at the box shown in Fig. 2 (g day^{-1}).

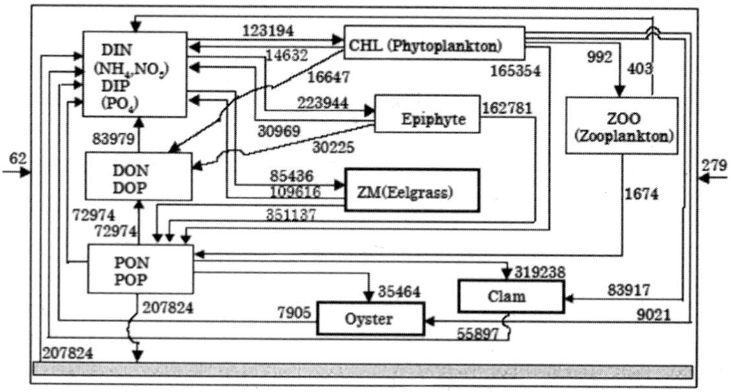

Figure 7: Daily averaged phosphorus flow among compartments at the box shown in Fig. 2 (g day^{-1}).

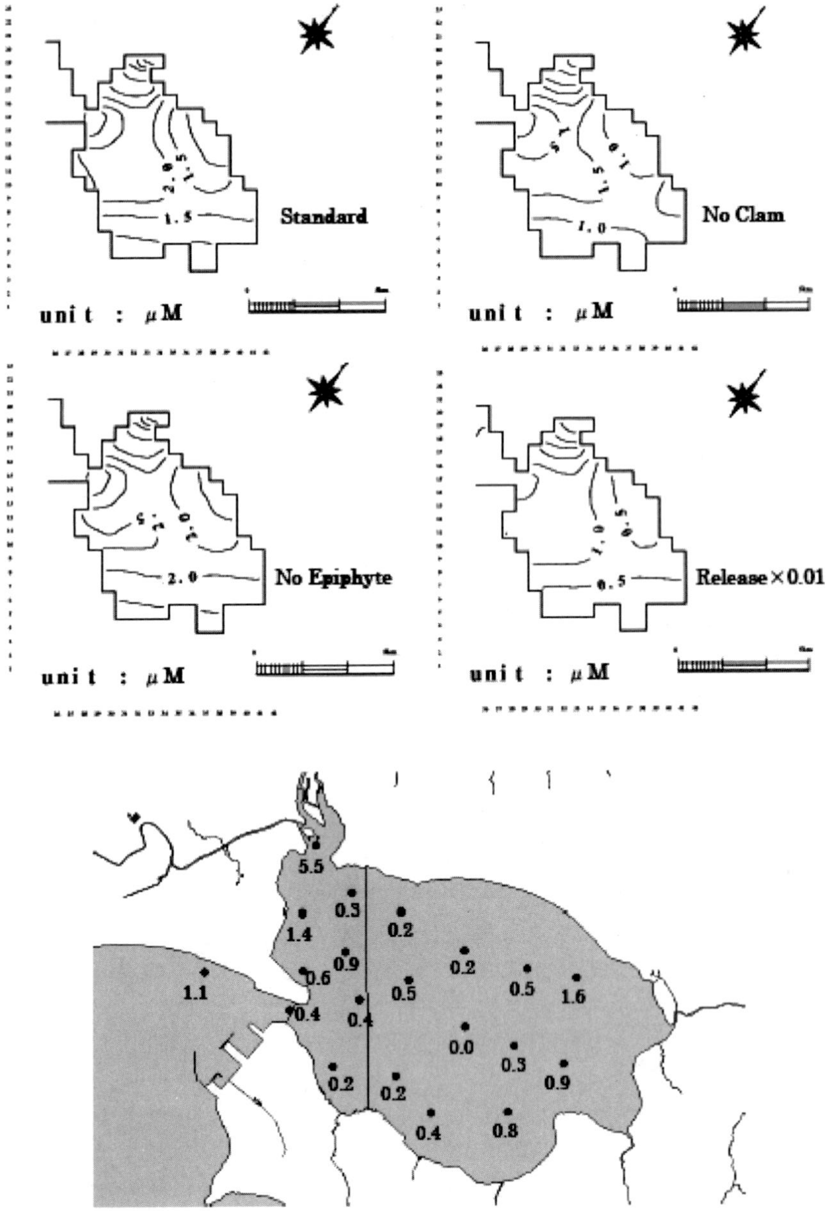

Figure 8: Simulated horizontal distribution at the surface of ammonium and observed one (bottom) (unit: micro mole l^{-1}). Four figures are: Standard: standard case, the nutrient release from the bottom corresponds to POM accumulation. No clam: without clam culture. No Epiphyte: without epiphyte. Release × 0.01: Nutrient release is 1% of POM accumulation (POM: particulate organic matter).

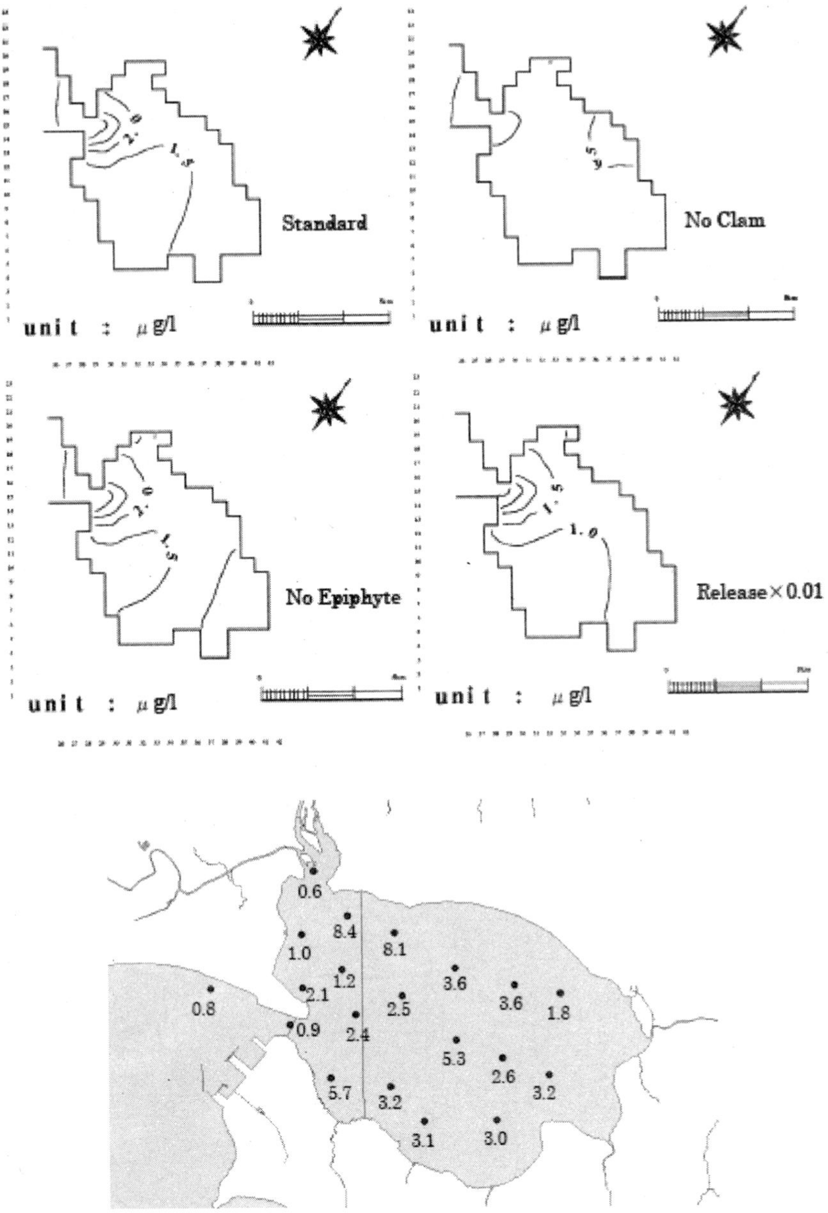

Figure 9: Simulated horizontal distribution at the surface of chlorophyll-a and observed one (bottom) (unit: micro g l^{-1}). Four figures are: Standard: standard case, the nutrient release from the bottom corresponds to POM accumulation. No clam: without clam culture. No Epiphyte: without epiphyte. Release × 0.01: Nutrient release is 1% of POM accumulation.

phytoplankton is lower at the south of the lake where the clam cultures are taking place. It is well known that shellfish acts as a filter in the marine ecosystem. The ammonium concentration is high at the river mouth (north-western part) due to the input from river and also is high at the southern lake, where the release from bottom mud is high due to the cumulative PON by fecal pellet of shellfish. In Figs 8 and 9, the results of sensitivity analysis are also described. In absence of clam, phytoplankton does not increase; however, in absence of epiphyte, there is a clear increasing trend. This suggests that epiphyte and phytoplankton is competing. In case with ammonium release smaller than PON accumulation (in the standard case, ammonium release is set to be same as PON sink), the simulated results of ammonium show a good agreement with the observed values. This suggests that PON accumulation is much higher than the nutrient release from the bottom. This phenomenon is characteristic in low temperature lagoons.

4. CONCLUSIONS

The following three important results are obtained from the numerical ecosystem-physical coupled model experiment for Akkeshi Lake. Clam plays most important role in the material cycle in Akkeshi Lake. Epiphyte is more important for shellfish food rather than phytoplankton. The amount of ammonium released from bottom mud is comparatively smaller than the accumulation of PON. The sensitivity analysis is one of the most important issues in numerical model studies especially for ecosystem model.

ACKNOWLEDGEMENTS

We wish to express our thanks to Dr. Pravakar Mishra of ICMAM for his valuable comments and help in proof reading of the manuscript.

REFERENCES

Iizumi, H., Akabane, H., Kishi, M.J. and Mukai, H., 2002. Effects of River Runoff on an Estuarine Ecosystem : Workshop Materials, The 3rd Joint Meeting of the Coastal Environmental Science and Technology Panel of UJNR.

Oshima, Y., Kishi, M.J. and Sugimoto, T., 1999. Evaluation of the nutrient budget in a seagrass bed *Ecological Modelling*, **115:** 19-33.

Winata, C.K., 2001. Population Study of the Manila clam (*Ruditapes philippinarum*) in Akkeshi-ko Estuary. A Dissertation presented to the Faculty of Science, Hokkaido University for the Degree of Doctor of Philosophy.

Wave Interaction with Floating and Submerged Rectangular Dykes in a Two-layer Fluid

P. Suresh Kumar, J. Bhattacharjee and T. Sahoo

Department of Ocean Engineering and Naval Architecture
Indian Institute of Technology Kharagpur, 721302, India
tsahoo1967@yahoo.com

1. INTRODUCTION

In recent years, there is a significant interest in the use of partial breakwaters to control waves. Most of these breakwaters are extended from the bottom up to the water surface, while partial breakwaters only occupy a segment of the whole water depth. In coastal engineering, partial barriers as breakwaters are more economical and sometimes more appropriate for engineering applications. These kinds of breakwaters also provide a less expensive means to protect beaches exposed to waves of small or moderate amplitudes, and to reduce the wave amplitude at resonance. A bottom-standing partial breakwater not only resists the wave propagation but also allows the navigation of vessels over it. The bottom-standing breakwaters are being used for fish farming in coastal fishery. In addition, these breakwaters create a calm region in the downstream of the wave motion and act as a sheltered region for a large group of marine habitats during severe wave conditions. Moreover, with the environmental concerns, the bottom-standing breakwater resists the sediment transport and provides a strong protection against coastal erosion. On the other hand, a surface-piercing breakwater does not require a strong bottom foundation and most suitable for protecting coastal and offshore structures in deep water region. The problems of propagation of water waves by floating/submerged obstacles have been studied theoretically by many investigators within the framework of linearized potential theory in a fluid domain of constant density. Some of the classical investigations on wave scattering by rigid obstacles are by Newman (1965), Miles (1967), Mei and Black (1969), Bai (1975) and McIver (1986). The interaction of linear water wave in a channel of constant depth impinging on a vertical thin porous breakwater with a semi-submerged and fixed rectangular obstacle in front of it is investigated by Yang et al. (1997).

Recently Söylemez and Gören (2003) studied the diffraction of oblique water waves by thick rectangular barriers. However, there is negligible progress on wave interaction with submerged or floating structures in two-layer fluid having a free surface and interface. Such situations arise in the ocean by solar heating of the upper layer, or in an estuary or a fjord into which fresh river water flows over oceanic water, which is more saline and consequently heavier. In such situations, the fluid can be idealized as a two-layer fluid by considering a lighter fluid of density ρ_1 lying over a heavier fluid of density ρ_2. It is observed that ships experience an abnormal resistance force in the Norwegian fjords, which was a mystery for a long time and was referred to as dead water, until Bjerknes explained that the reason behind this unknown phenomenon is due to the internal waves at the interface, which are generated by the motion of the ships (see Kundu and Cohen, 2002). The propagation of waves in a two-layer fluid (in the absence of any obstacles) was first investigated by Stokes (1847) and the classical problem of this type of two-layer fluid separated by a common interface with the upper fluid having a free surface is given in Lamb (1932) and Wehausen and Laitone (1960). In two-layer fluid, two types of wave modes exist because of the existence of free surface and interface. The wave generated due to the free surface are known as surface waves (propagating faster with a higher wave mode referred to as surface mode (SM)), and the waves generated due to the existence of the interface are known as internal waves (propagating slowly with a lower wave mode referred to as internal mode (IM)) (see Milne-Thomson, 1996). Linton and Mciver (1995) developed a general theory for a two-dimensional wave scattering by the horizontal cylinders in an infinitely deep two-layer fluid, and calculated the amount of energy that was converted from one wave number to the other for the case of circular cylinders in either the upper or lower fluid layer. The motivation for their work came from a plan to build an underwater pipe bridge across one of the Norwegian fjords, bodies of water which typically consists of a layer of fresh water about 10 m thick on top of a very deep body of salt water.

In the present work, the scattering of water waves by floating and submerged dykes in a two-layer fluid is investigated in two-dimensions in the context of linearized theory of water waves. The dykes are of rectangular geometry and two-layer fluid is considered to be of finite depth. Matched eigenfunction expansion method is used to solve the boundary value problem and due to the flow discontinuity at the interface, the eigenfunctions involved have an integrable singularity at the interface. The orthonormal relation used in the present analysis is a generalization of the classical one corresponding to a single-layer fluid. The reflection coefficients and force amplitude are plotted and present results in two-layer fluid are compared with those existing in the literature for a single-layer fluid.

2. MATHEMATICAL FORMULATION

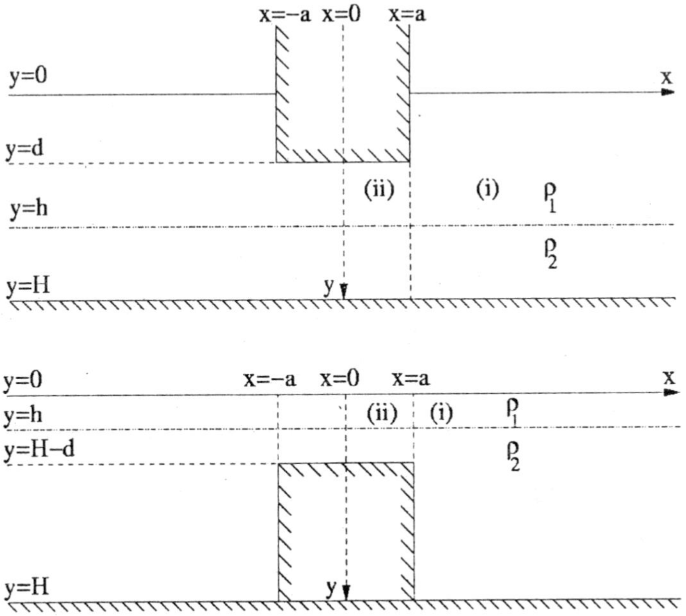

Figure 1: Definition sketch.

In the present study, the Cartesian co-ordinate system is chosen with x-axis in the direction of wave propagation and y-axis in the downward direction. The problem is considered in the two dimensions and the dykes of rectangular cross section are placed in the fluid domain as in Fig. 1. The fluid is assumed to be inviscid and incompressible and the wave motion is considered in the linearized theory of water waves by neglecting the effect of surface tension. In the two-layer fluid, the upper fluid has a free surface and the two fluids are separated by a common interface, each fluid is of infinite horizontal extent occupying the region $-\infty < x < \infty$; $0 < y < h$ in case of the upper fluid of density ρ_1, and $-\infty < x < \infty$; $h < y < H$ in case of the lower fluid of density ρ_1. The flow is assumed to be irrotational and simple harmonic in time and hence the velocity potential Φ (x, y, t) exists such that $\Phi\ (x, y, t) = \text{Re}\ [\phi_0\ \phi\ (x, y) \exp\ (-i\omega t)]$. The factor $\phi_0 = -igI_0/\omega$ is removed for convenience in the construction of eigenfunctions, where I_0 is the amplitude and ω is the wave frequency of the incident waves. The spatial velocity potential ϕ satisfies the Laplace equation in both the fluid regions and the linearized free surface boundary condition as given by

$$\frac{\partial \phi}{\partial y} + K\phi = 0 \text{ on } \quad y = 0, \tag{1}$$

where $K = \omega^2/g$, and g is the gravitational constant. At the interface, the continuity of the vertical component of velocity and pressure yield the boundary conditions

$$\left(\frac{\partial\phi}{\partial y}\right)_{y=h_+} = \left(\frac{\partial\phi}{\partial y}\right)_{y=h_-} \text{ and } \left(\frac{\partial\phi}{\partial y} + K\phi\right)_{y=h_+} = s\left(\frac{\partial\phi}{\partial y} + K\phi\right)_{y=h_-} \tag{2}$$

where $s = \rho_1/\rho_2$ with $0 < s < 1$. The condition on the solid boundaries is given by $\partial\phi/\partial n = 0$, where n is the outward normal to the boundary.

3. SOLUTION METHOD

3.1 Case of a Surface-Piercing Dyke

In this section, the wave scattering by a surface-piercing dyke as shown in Fig. 1 is discussed. The dyke has width $2a$ and wetted draft d. The symmetry of the configuration is being exploited to simplify the solution by considering ϕ (x, y) as a sum of symmetric and antisymmetric parts. Thus, ϕ is written as $\phi = \phi^s + \phi^a$, where the symmetric potential ϕ^s is an even function of x and the antisymmetric potential ϕ^a is an odd function of x. With this decomposition of velocity potential, the boundary value problem reduces to two simpler problems in the region $x \geq 0$ only.

Considering the waves incident from large positive x upon the dyke, the solution proceeds by taking eigenfunction expansions valid in each of the two regions marked in Fig. 1 and matching them on the common boundary. The symmetric velocity potentials ϕ^s in region (i) and region (ii) are given by

$$\phi_1^s = \begin{bmatrix} \frac{1}{2}\sum_{n=I}^{II} \exp\{p_n(x-a)\} + \\ \sum_{n=I,II,1}^{\infty} A_n^s \exp\{-p_n(x-a)\} \end{bmatrix} f_n(y), \text{ in region (i),} \tag{3}$$

$$\phi_2^s = B_0^s \chi_0(y) + \sum_{n=I,1}^{\infty} B_n^s \frac{\cosh \lambda_n x}{\cosh \lambda_n a} \chi_n(y), \text{ in region (ii).} \tag{4}$$

The corresponding expressions for the antisymmetric potential ϕ^a are

$$\phi_1^a = \begin{bmatrix} \frac{1}{2}\sum_{n=I}^{II} \exp\{p_n(x-a)\} \\ + \sum_{n=I,II,1}^{\infty} A_n^a \exp\{-p_n(x-a)\} \end{bmatrix} f_n(y), \text{ in region (i),} \tag{5}$$

$$\phi_2^a = B_0^a \frac{x}{a}\chi_0(y) + \sum_{n=I,1}^{\infty} B_n^a \frac{\sinh \lambda_n x}{\sinh \lambda_n a}\chi_n(y) \text{ in region (ii).} \tag{6}$$

In the open water region (i), the vertical eigenfunctions f_n's satisfying Laplace equation along with Eqs. (1-2) and condition on the solid boundary at seabed are given by

$$f_n(y) = \begin{cases} -N_n^{-1} \sin\{p_n(H-h)\} \dfrac{(p_n \cos p_n y - K \sin p_n y)}{p_n \sin p_n h + K \cos p_n h}, & 0 < y < h, \ (n = I, II, 1, 2, 3, \ldots) \\ N_n^{-1} \cos\{p_n(H-y)\}, & h < y < H. \end{cases} \tag{7}$$

where

$$\begin{aligned} N_n^2 = {} & (4p_n)^{-1}(p_n \sin p_n h + K \cos p_n h)^{-2}\big[s \sin^2 p_n(H-h) \\ & \{p_n^2(2p_n h + \sin 2p_n h) + K^2(2p_n h - \sin 2p_n h) + \\ & 2Kp_n(\cos 2p_n h - 1)\} + (p_n \sin p_n h + K \cos p_n h)^2 \\ & \{2p_n(H-h) + \sin 2p_n(H-h)\}\big] \end{aligned} \tag{8}$$

$p_I = -im_1$, $p_{II} = -im_2$, where m_n, $n = 1, 2$ are positive real and p_n, $n \geq 1$ are positive purely imaginary roots of the dispersion relation in p as given by

$$\begin{aligned} & (1-s)p^2 \tanh p(H-h) \tanh ph - pK\{\tanh ph + \tanh p(H-h)\} \\ & + K^2\{s \tanh ph \tanh p(H-h) + 1\} = 0. \end{aligned} \tag{9}$$

Here m_1 and m_2 represent the propagating modes for the surface and internal waves respectively. For the dyke covered region (ii), the eigenfunctions satisfying Laplace equation along with Eq. (2), and the condition on the solid boundary on the seabed ($y = H$) and at the bottom of the dyke ($y = d$), are given by

$$\chi_0(y) = \begin{cases} L_0^{-1}, & d < y < h, \\ sL_0^{-1}, & h < y < H, \end{cases}$$

$$\chi_n(y) = \begin{cases} -L_n^{-1} \sin \lambda_n(H-h) \dfrac{\cos \lambda_n(y-d)}{\sin \lambda_n(h-d)}, & d < y < h, \ n = I, 1, 2, \ldots \\ L_n^{-1} \cos \lambda_n(H-y), & h < y < H, \end{cases} \tag{10}$$

where $L_0^2 = s\{s(H-h)+(h-d)\}$

and

$$L_n^2 = (4\lambda_n)^{-1}\{\sin\lambda_n(h-d)\}^{-2} \times [s\sin^2\lambda_n(H-h)\{2\lambda_n(h-d) + \sin 2\lambda_n(h-d)\} + \sin^2\lambda_n(h-d)\{2(H-h)\lambda_n + \sin 2\lambda_n(H-h)\}] \quad (11)$$

$\lambda_I = -i\beta$ where β is positive real and λ_n, $n \geq 1$ are positive purely imaginary roots of the dispersion relation in λ as given by

$$(1-s)\lambda\tanh\lambda(H-h)\tanh\lambda(h-d) - K\left\{\begin{array}{l}\tanh\lambda(h-d)\\ +s\tanh\lambda(H-h)\end{array}\right\} = 0. \quad (12)$$

It may be noted that in the dyke covered region, in case of $d < h$ the vertical eigenfunctions have only one propagating mode because of the presence of the interface, whereas in the case of $d \geq h$ the term $n = I$ does not appear in the expression of eigenfunctions and the corresponding vertical eigenfunctions becomes

$$\chi_n(y) = L_n^{-1}\cos\lambda_n(H-y),\ n = 0,1,2,\ldots, \quad (13)$$

$L_0^2 = 1 - d/H$ and $L_n^2 = 0.5(1-d/H)$, for $(n \geq 1)$,

$$\text{where } \lambda_n = n\pi/(H-d),\ n = 0,1,2,\ldots. \quad (14)$$

The aforementioned expressions for ϕ^s and ϕ^a satisfy all of the boundary conditions except those on $x = a$. It may be noted that $A_n^{s,a} = 0.5R_n^{s,a}$ for $n = I, II$, where $R_I^{s,a}$ and $R_{II}^{s,a}$ are related with the reflection coefficients Kr_I and Kr_{II} in surface and internal modes respectively. It may be noted that the aforementioned eigenfunctions are orthonormal with respect to the inner products as given below

$$\langle f_n, f_m\rangle_1 = s\int_0^h f_n f_m\,dy + \int_h^H f_n f_m\,dy,$$

$$\langle \chi_n, \chi_m\rangle_2 = \begin{cases} s\int_d^h \chi_n\chi_m\,dy + \int_h^H \chi_n\chi_m\,dy, & d < h, \\ \int_d^H \chi_n\chi_m\,dy, & d \geq h. \end{cases} \quad (15)$$

The unknowns $\{A_n^s, A_n^a,\ n = I, II, 1, \ldots\}$ and $\{B_n^s, B_n^a,\ n = 0, I, 1, \ldots\}$ are determined by using the continuity of pressure and horizontal velocity at $x = a$ along with the suitable utilization of the orthonormal relations as defined in the Eq. (15).

3.2 Case of a Bottom-Standing Dyke

The scattering of incident wave trains by a bottom-standing dyke in the two-layer fluid is explained in the present section. The symmetricity of the problem is again exploited by writing the solution as the sum of a symmetric and an antisymmetric part. The line of symmetry is same as before and the origin of x has been chosen to be at the centre of the dyke (Fig. 1). The symmetric and antisymmetric parts of the velocity potential in the region (i) is same as that described in case of a surface-piercing dyke. The symmetric and antisymmetric parts of the velocity potential in region (ii) are as given below.

$$\phi_2^s = \sum_{n=I,\,II,\,1}^{\infty} B_n^s \frac{\cosh \lambda_n x}{\cosh \lambda_n a} \chi_n(y), \quad \phi_2^a = \sum_{n=I,\,II,\,1}^{\infty} B_n^a \frac{\sinh \lambda_n x}{\sinh \lambda_n a} \chi_n(y). \tag{16}$$

The expression of $\chi_n(y)$ is very similar to that of open region vertical eigenfunction f_n as defined in Eq. (7) and λ_n satisfy a dispersion relation similar to that given in Eq. (9). All other solution procedures in the case of bottom-standing dyke are very similar to that used in the case of surface-piercing dyke.

4. NUMERICAL RESULTS AND DISCUSSION

Numerical results are computed and analyzed for the wave scattering by surface-piercing and bottom-standing dykes in a two-layer fluid. The effects of various non-dimensional physical parameters on wave reflection in both SM and IM are analyzed. For convenience, the wave parameters are given in terms of the non-dimensional wave number $m_1 d$, water depth h/H, fluid density ratio s and dyke parameters a/d, H/d and b/H.

4.1 Case of a Surface-Piercing Dyke

In the present subsection, the reflection coefficients in SM and IM and the hydrodynamic forces on the surface-piercing dyke are analyzed. For sake of simplicity, all the results are plotted with respect to the normalized wave number $m_1 d$ in SM.

In Fig. 2, the reflection coefficients Kr_I and Kr_{II} in a two-layer fluid for two different values of H/d ratios are compared with the results obtained by Mei and Black (1969) in a single-layer fluid. In general, the reflection coefficients in SM are similar to that observed by Mei and Black (1969)

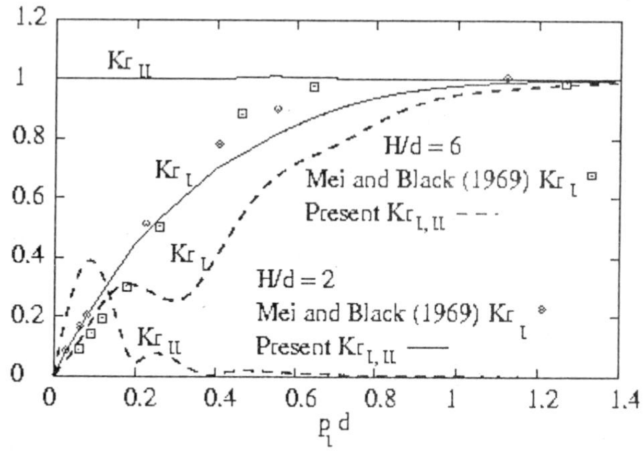

Figure 2: Comparison of reflection coefficients in SM, Kr_{I} and IM, Kr_{II} versus m_1d for two different H/d values at $a/d = 1.0$, $h/H = 0.25$ and $s = 0.75$ with Mei and Black (1969).

except in case of intermediate frequency range, where Kr_{I} is found to be significantly small. It may be noted that in case of a two-layer fluid, due to the presence of the interface, waves in SM and IM propagate below the dyke (when $d < h$), which is not the case for single-layer fluid. For small values of m_1d, which corresponds to long wave region, the interface is very close to the dyke and allows more wave to reflect by the structure. Whilst, for large values of m_1d, which corresponds to short wave region, the interface is far from the free surface and wave transmission in SM due to interface becomes insignificant. On the other hand, for intermediate frequency range, the waves in SM transmitted significantly due to the presence of the interface and this may be the reason for smaller reflection coefficients in SM as observed in the present study as compared to the wave reflection in the single-layer fluid. When the surface dyke is above the interface, the general trend of the IM wave reflection coefficients observed in the two-layer fluid is found to be similar to the one observed for a bottom-standing dyke in a single-layer fluid (see Mei and Black (1969); Fig. 2). However, when the surface dyke touches the interface or it is extended beyond the interface the reflection coefficient in IM is found to be 100%.

The variation of reflection coefficients in SM and IM versus m_1d are plotted in Fig. 3 for different values of H/d. For all values of H/d with an increase in m_1d the wave reflection in SM increases and attains a 100% reflection in the deepwater region. On the other hand, for $d/H < h/H$, the general trend of reflection coefficient in IM follows an oscillating pattern and it attains a zero reflection in the deepwater region. This is due to the fact that in the deepwater region, the dyke is far from the interface and hence it has a negligible impact on waves in IM. However, as the dyke approaches towards the interface, the wave reflection in IM increases sharply and attains a 100% reflection over the entire frequency range in case of $d \geq h$.

The effect of dyke width to wetted draft ratio, a/d on reflection coefficients in SM and IM are shown in Fig. 4. It is observed that with an increase in a/d ratio, the reflection in both SM and IM increases. This is expected because an increase in dyke width will enhance the wave reflection and when the width becomes infinitely large, the wave reflection in SM become 100% over entire frequency range because in such situation there will not be any transmitted wave in SM existing in the downstream direction. On the other hand, the wave reflection in IM is similar to the one observed in the case of a bottom-standing dyke in a single-layer fluid (see Fig. 2 of MEI Mei and Black, 1969).

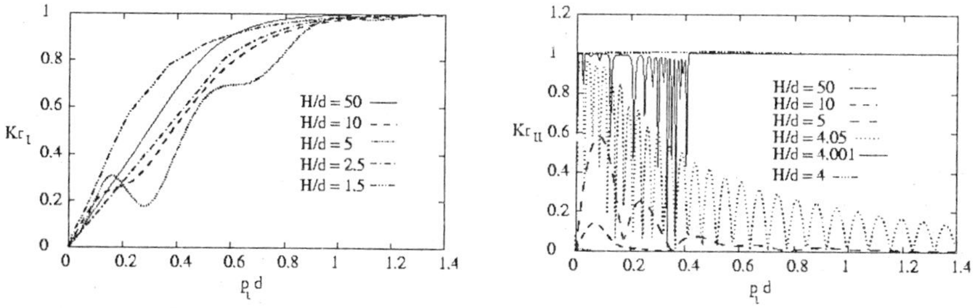

Figure 3: Reflection coefficients in SM, Kr_{I} and IM, Kr_{II} versus m_1d for different H/d values at $a/d = 1.0$, $h/H = 0.25$ and $s = 0.75$.

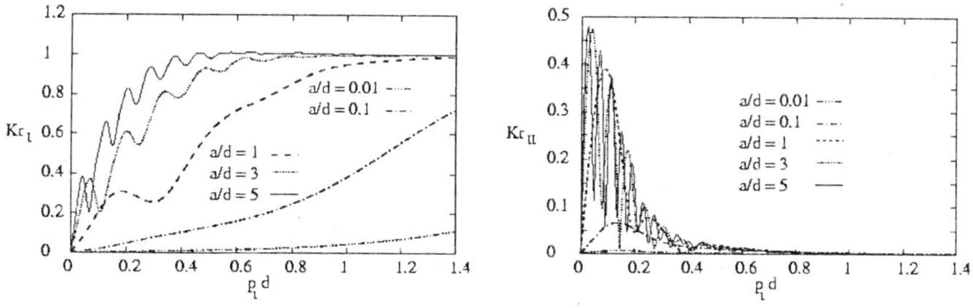

Figure 4: Reflection coefficients in SM, Kr_I and IM, Kr_{II} versus m_1d for different a/d values at $H/d = 6.0$, $h/H = 0.25$ and $s = 0.75$.

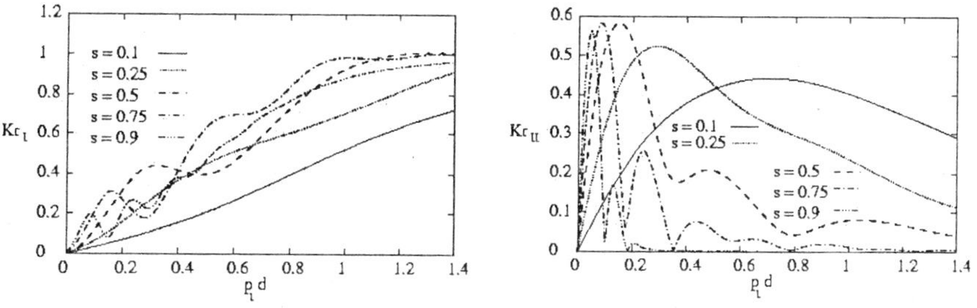

Figure 5: Reflection coefficients in SM, Kr_I and IM, Kr_{II} versus m_1d for different s values at $H/d = 5.0$, $a/d = 1.0$ and $h/H = 0.25$.

Reflection coefficient is plotted versus $m_1 d$ in SM and IM for various values of s in Fig. 5. It is observed that in the deepwater region, the wave reflection in SM increases with an increase in the value of s and a reverse trend is observed in case of IM wave reflection.

The magnitude of horizontal and vertical hydrodynamic forces per unit incident wave amplitude and length of dyke are analysed in Figs 6 and 7. The pattern of hydrodynamic forces for single cylinder in case of the two-layer fluid is similar to the one explained in Mciver (1985) in case of a pair of cylinder in the single-layer fluid. In addition, resonance behaviour is observed in the case of horizontal hydrodynamic force. As a result, a rise in the horizontal hydrodynamic force at resonance frequency will lead to a higher drag on the obstacle. However, it is observed that the magnitude of resonating horizontal hydrodynamic force increases with a decrease in H/d value and an increase in a/d value. On the other hand, the vertical hydrodynamic force increases with an increase in $m_1 a$ and in the deep water region the magnitude of vertical hydrodynamic force is high for small H/d and a/d values.

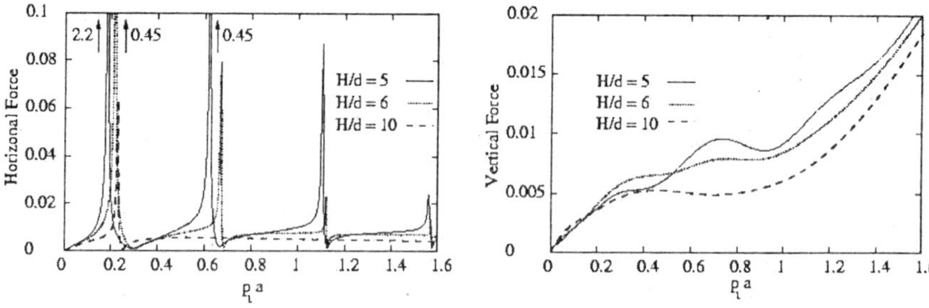

Figure 6: Horizontal and vertical force per unit incident wave amplitude and length of dyke in MN/m^2 for different H/d values at a/d = 1.0, s = 0.75 and h/H = 0.25.

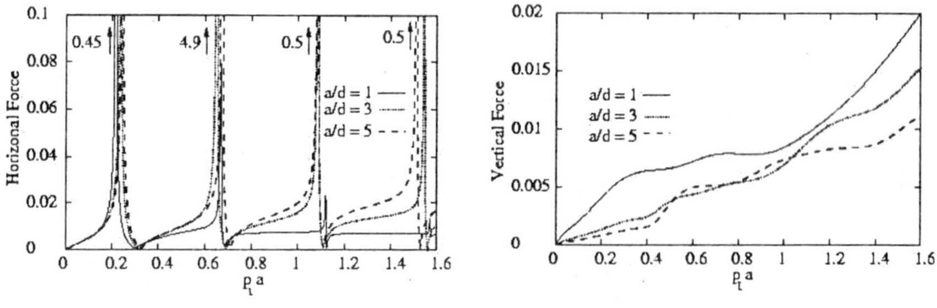

Figure 7: Horizontal and vertical force per unit incident wave amplitude and length of dyke in MN/m^2 for different a/d values at H/d = 6.0, s = 0.75 and h/H = 0.25.

4.2 Case of a Bottom-Standing Dyke

In the present section, the reflection coefficients in SM and IM for the bottom-standing dyke are analyzed for various physical parameters of interest.

The variation of reflection coefficients in SM and IM versus $m_1 d$ are plotted in Fig. 8 for different values of H/d. In general, it is observed that the wave reflection in both SM and IM are found to be increasing with a decrease in H/d. When the bottom-standing dyke is in the lower fluid domain $(H - d)/H > h/H$, i.e., $H/d = 2.0$ and 3.0), with an increase in $m_1 d$ the wave reflection in both SM and IM decreases and approaches to zero in the deepwater region. This is because the bottom-standing dyke has a negligible impact on the wave motion in both the modes in the deepwater region. On the other hand, when the dyke is extended up to the upper fluid ($(H - d)/H < h/H$, i.e., $H/d = 1.05$ and 1.15), the wave reflection in SM increases up to certain intermediate frequency range; thereafter it starts decreasing and the trend suggests that it attains a zero reflection in the deep water region. The increase in the wave reflection in SM, for certain intermediate frequency range in case of $(H - d)/H < h/H$ is due to the fact that the waves propagating at the interface get completely blocked in such situations and this will lead to a higher wave reflection. For $(H - d)/H < h/H$, the wave reflection in IM is found to be increasing with an increase in $m_1 d$ and the trend suggests that

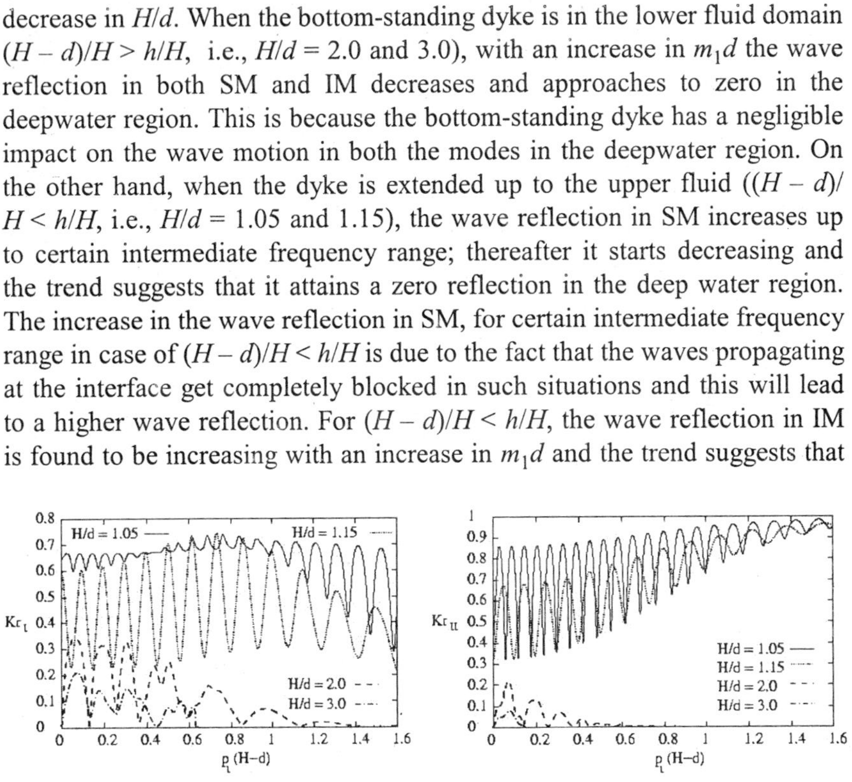

Figure 8: Reflection coefficients in SM, Kr_I and IM, Kr_{II} versus $m_1 d$ for different H/d values at $a/d = 6.0$, $h/H = 0.25$ and $s = 0.75$.

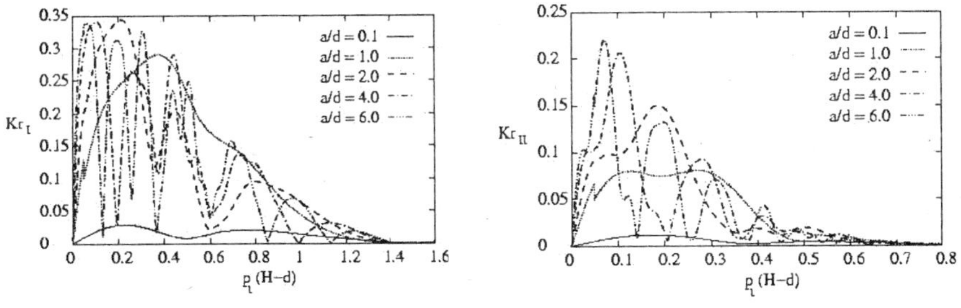

Figure 9: Reflection coefficients in SM, Kr_I and IM, Kr_{II} versus $m_1 d$ for different a/d values at $H/d = 2.0$, $a/d = 0.25$ and $h/H = 0.75$.

the reflection in IM attains 100% reflection in the deepwater region. This is because in deep water the interface is far from the free surface and in such situation free surface cannot transmit the waves in IM and this will lead to a situation where there will be no wave transmission in IM.

The effect of dyke width to wetted draft ratio, a/d on reflection coefficients in SM and IM are shown in Fig. 9. It is observed that with an increase in a/d ratio, the reflection in both SM and IM increases. The general trend of the reflection coefficient pattern in both SM and IM is similar to the one observed in the case of a bottom-standing dyke in a single-layer fluid (see Fig. 2 of Mei and Black, 1969).

The effect of interface position h/H on reflection coefficients in SM and IM are shown in Fig. 10. It is observed that for both $(H - d)/H > h/H$ and $(H - d)/H < h/H$, with a decrease in h/H ratio, the reflection in SM increases. However the general trend of the reflection coefficient variation in SM is different for $(H - d)/H > h/H$, and $(H - d)/H < h/H$. On the other hand, the wave reflection in IM increases with an increase in the value of h/H.

Reflection coefficients are plotted versus $m_1 d$ in SM and IM for various values of s in Fig. 11. It is observed that the wave reflection in SM has higher reflection peaks for higher values of s and a reverse trend is observed in case of IM wave reflection.

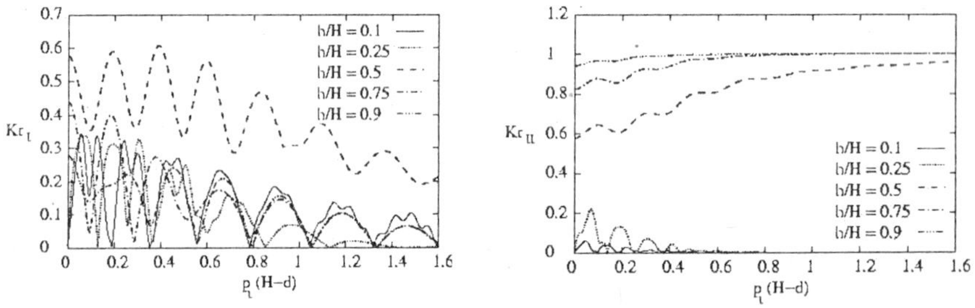

Figure 10: Reflection coefficients in SM, Kr_{I} and IM, Kr_{II} versus $m_1 d$ for different h/H values at $H/d = 2.0$, $a/d = 6.0$ and $s = 0.75$.

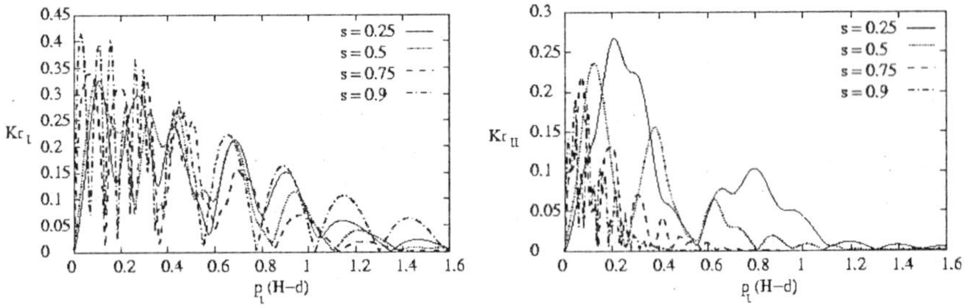

Figure 11: Reflection coefficients in SM, Kr_{I} and IM, Kr_{II} versus $m_1 d$ for different s values at $H/d = 2.0$, $a/d = 6.0$ and $h/H = 0.25$.

5. CONCLUSIONS

The wave scattering by rectangular surface-piercing and bottom-standing dykes in a two-layer fluid is investigated. Orthogonal relations suitable for the two-layer fluid, in both the cases of open water region and dyke covered region are utilized to solve the problems. The symmetricity of the problem is exploited by splitting the velocity potential in symmetric and antisymmetric parts and this made the problem simple and the number of unknowns to be found out by solving the matrix system are also reduced dramatically. The wave reflection for both surface-piercing and bottom-standing dykes is found to be strongly dependent on the interface location and the fluid density ratio apart from the dyke geometry. Moreover, the hydrodynamic force on the surface-piercing dyke is also studied and a strong resonating horizontal force is observed for certain range of frequencies. A similar approach can be utilized to study more general problems in two-layer fluid having a free surface.

ACKNOWLEDGEMENTS

J. Bhattacharjee acknowledges CSIR, New Delhi for the financial support received in terms of Junior Research fellowship.

REFERENCES

Bai, K.J., 1975. Diffraction of oblique waves by an infinite cylinder. *J. Fluid Mech.*, **68:** 513-535.

Kundu, P.K. and Cohen, I.M., 2002. *Fluid Mechanics* (2nd ed.). Academic Press, San Diego, California, USA.

Lamb, H., 1932. *Hydrodynamics* (6th ed.). Cambridge University Press. Reprinted 1993.

Linton, C.M. and Mciver, M., 1995. The interaction of waves with horizontal cylinders in two-layer fluids. *J. Fluid Mech.*, **304:** 213-229.

Mciver, P., 1986. Wave forces on adjacent floating bridges. *J. Appl., Ocean Res.*, **8(2):** 67-75.

Mei, Chiang C. and Black, J.L., 1969. Scattering of surface waves by rectangular obstacles in waters of finite depth. *J. Fluid Mech.*, **38:** 499-511.

Miles, J.W., 1967. Surface-wave scattering matrix for self. *J. Fluid Mech.*, **28:** 755-767.

Milne-Thomson, L.M., 1996. *Theoretical Hydrodynamics*. Dover Publications, Inc., New York.

Newman, J.N., 1965. Propagation of water waves past long two-dimensional obstacles. *J. Fluid Mech.*, **23:** 23-29.

Söylemez, M. and Gören, Ö., 2003. Diffraction of oblique waves by thick rectangular barriers. *J. Appl., Ocean Res.*, **25:** 345-353.

Stokes, G.G., 1847. On the theory of oscillatory waves. *Trans. Camb. Phil. Soc.*, **8:** 441-455. Reprinted in *Maths. Phys. Papers*, Cambridge University Press., **1:** 314-326.

Wehausen, J.V. and Laitone, E.V., 1960. Handbuch der physik. (Ed. S. Flugge). Springer-Verlag, **9:** 446-778.

Yang, H.T.Y., Huang, L.H. and Hwang, W.S., 1997. The interaction of a semi-submerged obstacle on the porous breakwater. *J. Appl., Ocean Res.*, **19:** 263-273.

Reef—An Ecofriendly and Cost Effective Hard Option for Coastal Conservation

Kiran G. Shirlal, Subba Rao, Radheshyam B. and Venkata Ganesh

Department of Applied Mechanics and Hydraulics
National Institute of Technology, Karnataka
Surathkal, Srinivasnagar - 575025, India

1. INTRODUCTION

Rubble mound breakwaters are the structures which are meant to reflect and dissipate energy of the wind generated waves and thereby to prevent their incidence on water area intended to protect. Submerged breakwater with its crest at or below still water level (SWL) can cause substantial wave attenuation and can be effectively used in places where tidal variations are small and only partial protection from waves is required, like harbour entrance, beach protection, small craft harbours etc.

The wave breaking over submerged breakwater causes great turbulence on lee side. Current and turbulence together on lee side of submerged breakwater have a strong power of erosion on a sandy bottom and can thus prevent siltation. They also offer resistance through friction and turbulence created by breakwater interference in wave field causing maximum wave damping and energy dissipation, minimum wave reflection and bottom scour, and maximum sand trapping efficiency (Baba, 1985; Pilarczyk and Zeilder, 1996). They are also used for coastal protection.

The reef is a structure which is little more than a homogeneous pile of stones without a layered structure. The hydrodynamic performance of the reef is investigated based upon physical model study to ascertain its suitability as coastal defense structure. The varying geometry and seaward location and wave transmission at the reef will help in designing an optimum structure.

2. LITERATURE REVIEW

The influence of the slope, crest width and depth of submergence of various shapes of submerged breakwaters on wave transmission was studied by

Johnson et al., 1951; Dattatri, 1978; Khader and Rai, 1980; Dick and Brebner, 1968; Smith et al., 1996; Pilarczyk and Zielder, 1996; Twu et al., 2001; Shirlal and Rao, 2003 and Shirlal et al., 2003. Some of the above authors opined that the submerged structure is constructed in a water depth of 1.5 m to 5 m with a slope of 1:2 to 1:3 and a height exceeding 0.7 times the depth of water. A reinforced concrete smooth submerged breakwater experimented in Russia with a seaward slope of 1:1.67 and vertical shoreward slope gives optimum wave transmission with minimum reflection for a tidal range less than 2 m and steepness greater than 0.075 (Baba, 1985). But there are as many opinions as the number of investigators on what should be the crest width of the submerged breakwater. In case of submerged structure the wave attacks on its crest and less on the seaward slope. Hence slope angle is not the governing parameter for stability. Various investigators have tested reefs of stone armour with steeper slopes of 1:1.5, 1:1.67 and 1: 1.75. However, better dissipation of waves, lower reflection and easier transport of sediment over the structure were observed for submerged breakwaters with seaward slopes ranging from 1:2 to 1:3 (Pilarczyk and Zielder, 1996). Ahrens (1984) showed that for a submerged reef, wave reflection is less than 20% for slopes of 1:1.67 for zero freeboard which is critical.

The reef is a structure which is little more than a homogeneous pile of stones whose weight is sufficient to resist the wave attack. A submerged reef is an optimized structure to highest degree. The reef is fundamentally built to break the steep waves and as this structure is submerged and porous, wave reflection is small and wave energy damping and wave transmission are the significant characteristics. The important reef parameters affecting the wave breaking and transmission are structure height, crest width and submergence. Therefore, an experimental investigation was carried out to determine the impact of the above parameters of submerged reef breakwater on wave height attenuation and wave transmission. Ahrens (1984 and 1989), Gadre et al. (1992), Pilarczyk and Zeilder (1996) and Nizam and Yowono (1996) have presented equations and graphs to calculate the armour weight of submerged reef breakwater.

3. DETAILS OF MODEL SETUP

3.1 Wave Flume

Physical model studies are conducted in a two-dimensional wave flume of 50 m × 0.71 m × 1.1 m in which regular waves are generated. It has a smooth concrete bed for a length of 42 m as shown in Fig. 1. The flume has bottom hinged flap type wave generator operated by a 7.5 HP, 11 KW, 1450 rpm induction motor. This motor is regulated by an inverter drive (0-50 Hertz) rotating at 0-155 rpm. The system can generate waves of 0.02 m to 0.24 m of 0.8 sec to 4 sec period in a maximum water depth (d) of 0.5 m.

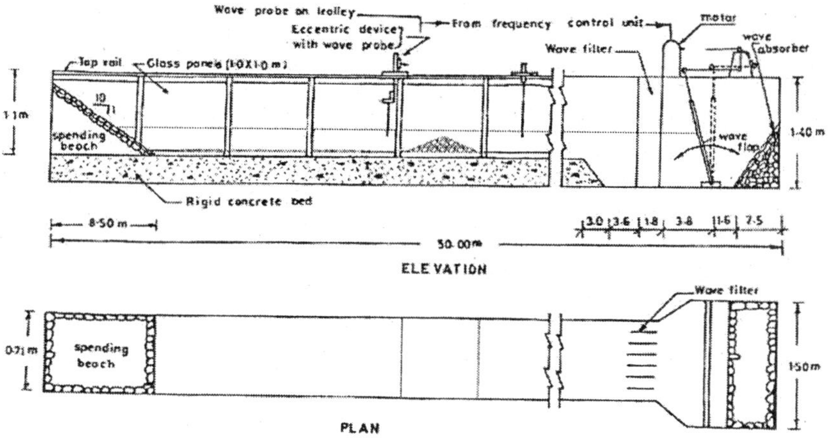

Figure 1: Details of experimental setup.

3.2 Reef Model

The 1:30 scale model of a reef of height (*h*) 0.25 m has a crest width (*B*). The reef is constructed with armour of weight W_{50} gms. The structure is of uniform slope of 1V:2H and armour units have a fitted placement. The crest width and armour weight are varied depending upon the requirements during the course of the experiment.

3.3 Instrumentation and Data Acquisition

The capacitance type wave probes along with amplification units are used for data acquisition. Two such probes are used during the experimental work, one for acquiring incident wave height (H_i) and the other for transmitted wave height (H_t). During the experimentation, the signals from wave channels are verified with digital oscilloscope along with computer data acquisition system. The wave probes are calibrated at the beginning of the work. Before starting the experiment, the flume is calibrated with breakwater model in place for different water depths to find out the incident wave heights for different combinations of wave height and wave period. Combinations that produced the secondary waves in the flume are not considered for the experiments. One probe is positioned at 1 m seaward of the reef and another at the locations where the transmitted wave height is to be measured. The signals from the wave probe are recorded for the incident and transmitted wave. Incident and transmitted wave heights are also measured manually as a crosscheck.

4. EXPERIMENTAL PROCEDURE

4.1 Study of Reef Stability

The 1V:2H sloped trapezoidal reef of height (h) 0.25 m with a crest width (B) of 0.1 m is constructed over the flat bed of the flume with armour of weight varying from 15 gms to 35 gms as the various design criteria given by Ahrens (1984 and 1989), Gadre et al. (1992), Nizam and Yuwono (1996) and Piarczyk and Zeidler (1996). This test section is subjected to normal wave attack of regular waves of height ranging from 0.1 m to 0.16 m of periods varying from 1.5 sec to 2.5 sec in a depth of water of 0.3 m as the reef stability is critical at the lowest water level.

4.2 Study of Wave Transmission at the Reef

The wave flume is filled with ordinary tap water to the required depths (d) of 0.3 m, 0.35 m and 0.4 m height. In the present model study, rigid bed conditions are considered and it is assumed that the onshore and offshore movement of sediments does not interfere in the wave attenuation process. The model is subjected to normal regular wave attack of height (H_i) of 0.10 m, 0.12 m, 0.14 m and 0.16 m with varying periods (T) of 1.5 sec, 2.0 sec and 2.5 sec in different water depths (d) mentioned above. The transmitted wave heights (H_t) on the leeside are observed for a distance of every metre up to 8 m i.e. at X/d ranges of 2.5 to 26.67.

5. RESULTS AND DISCUSSIONS

The damage of the reef and wave height attenuation (i.e. WHA = 1 – (H_i/H_t)) due to reef, is investigated with respect to incident wave steepness parameter and reef crest width.

5.1 Stability of the Reef

The damage to the reef, of crest width B of 0.1 m, height h of 0.25 m and with armour stones of weight varying from 15 gms to 35 gms, for a critical depth of 0.3 m is recorded in the form of reduction in crest height (h_c). The dimensionless damage is computed as h_c/h. The variation of the dimensionless damage h_c/h with spectral stability number N_S* for varying armour stone weight is shown in Fig. 2. The spectral stability number N_S* (Ahrens, 1984) is given as

$$N_S{}^* = \frac{H(S)^{-1/3}}{\Delta D_{n50}} \tag{1}$$

H is the design wave height, S is the local wave steepness, Δ is mass density of armour stone and D_{n50} is the nominal diameter of the armour stone. From the figure, it is observed that a reef with armour stones of weight of 35 gms

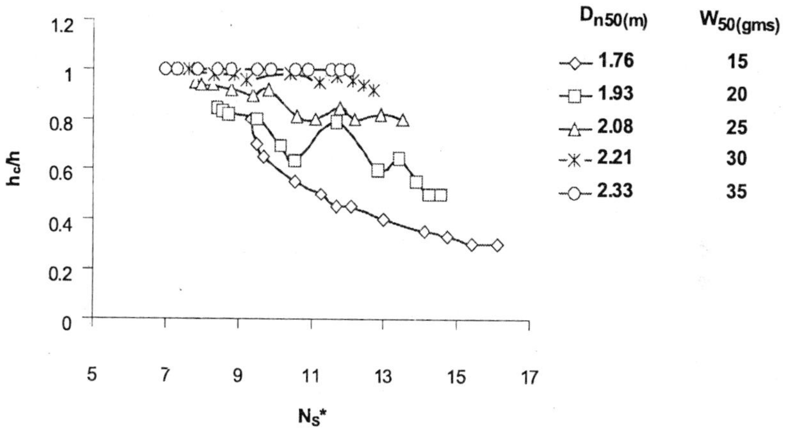

Figure 2: Damage (h_c/h) of the reef with spectral stability number (N_S*).

is stable while armour stones of 30 gms are also quite stable. Hence, armour stone of 30 gms is chosen as optimum weight for a stable reef.

5.2 Impact of Reef Location on Wave Height Attenuation

The submerged reef successfully trips the steeper waves and dissipates wave energy (Fig. 3). The effectiveness of reef in damping of waves increases with an increase in wave steepness. Therefore, reef can be successfully used for coastal protection where they are subjected to steep waves during storms. Further, as the distance (X) increases, the waves that break over the reef, loose some more energy while propagating in the stilling basin (i.e. the energy dissipation zone). The transmitted wave heights on the leeside are

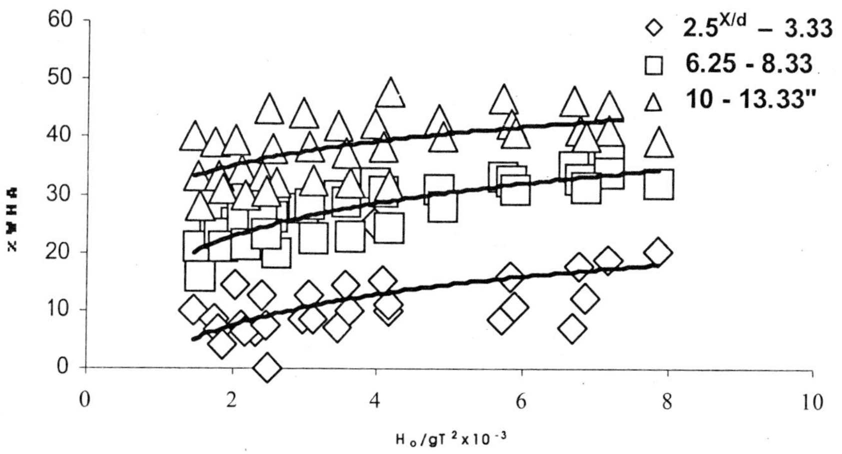

Figure 3: Variation of WHA with the deep water wave steepness parameter for varying reef locations.

observed for a distance (X/d) varying from 2.5 to 26.67. It is observed that up to a shoreward distance (X/d) of 10.0 to 13.33, there is a maximum wave attenuation of about 50% beyond which there is no significant increase in wave attenuation. This is the maximum wave height attenuation observed in sea at natural bar according to Battjes and Janssen as quoted by Cox and Clark (1992). Therefore, it is decided to locate the reef within a maximum shoreward distance (X/d) of 13.33 for optimizing the location of the reef and its crest width. The trend lines show that for X/d = 2.5 to 3.33, the waves are attenuated by a maximum amount of 18%, for X/d = 6.25 to 8.33, the maximum attenuation of wave heights is about 33% and for X/d = 10 to 13.33, it is about 43%. Therefore, to study the effect of the reef crest width, the wave height attenuation is recorded for X/d of 6.25 to 8.33.

Figure 4 shows the trends of wave height attenuation (WHA), for a reef at a shoreward distance (X/d) of 6.25 to 8.33, against varying deep water wave steepness parameter (H_0/gT^2) for a range of crest widths (B/d = 0.25 to 1.33). For a given crest width the WHA increases with an increase in steepness parameter. The reason is the reef breaks the steeper waves successfully, increasing wave damping, resulting in increased WHA which finally reduces wave transmission. The WHA decreases with an increase in B/d. This is because after breaking, wider reef offer friction and waves increasingly shoal over them. The maximum WHA are 33%, 37%, 41% and 46% for the ranges of crest widths (B/d) from 0.25 to 0.33, 0.5 to 0.67, 0.75 to 1.0 and 1.0 to 1.33 respectively. For the reef of crest width (B/d) of 1 to 1.33, the wave height attenuation at X/d = 6.25 to 8.33, is 46% which is

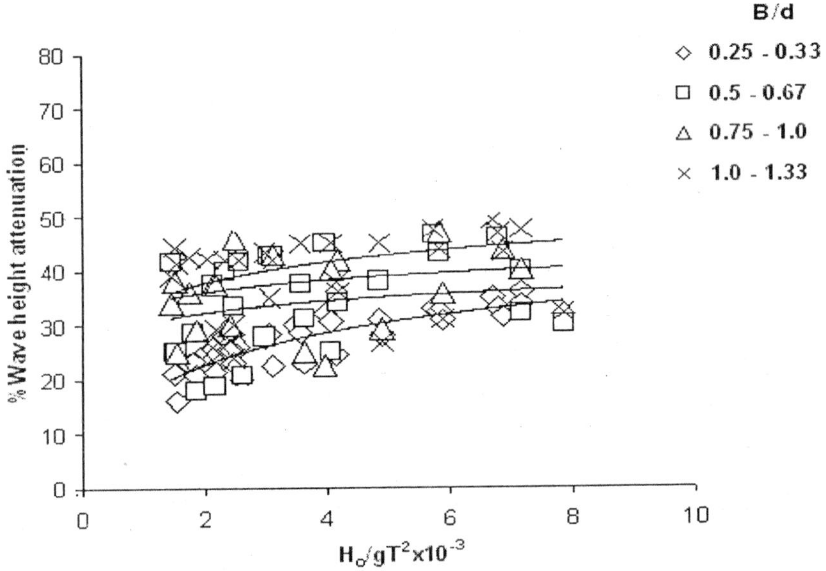

Figure 4: Variation of WHA with deep water wave steepness parameter for varying reef crest widths.

more than that (43%) achieved by the reef of crest width (*B*/*d*) of 0.25 to 0.33, at *X*/*d* = 10 to 13.33.

6. CONCLUSIONS

The following conclusions are drawn from the present study:

- Optimum armour weight of the submerged reef is 30 gms.
- Wave height attenuation for the submerged reef increases with an increase in H_0/gT^2, *X*/*d* and *B*/*d*.
- For the reef of crest width (*B*/*d*) of 0.25 to 0.33, the maximum wave height attenuation of wave heights is about 18% to 43% at *X*/*d* varying from 2.5 to 13.33.
- For the reef of crest width (*B*/*d*) of 1 to 1.33, located at *X*/*d* = 6.25 to 8.33, the wave height attenuation is 46% which is more than that achieved by the reef of crest width (*B*/*d*) of 0.25 to 0.33, located at *X*/*d* = 10 to 13.33.
- The submerged reef, which is an optimised structure, attenuates the steep waves; may be an optimum coastal protection structure.

ACKNOWLEDGEMENTS

The authors are thankful to the Director of National Institute of Technology Karnataka, Surathkal, and Head of the Department of Applied Mechanics and Hydraulics for the facilities provided for the investigation and permission granted to publish the results.

REFERENCES

Abdul Khade, M.H. and Rai, S.P., 1980. A study of submerged breakwaters. *Journal of Hydraulic Research*, **28(2):** 113-121.

Ahrens, J.P., 1984. Reef type breakwaters. Proceedings of 19th Coastal Engineering Conference. ASCE, pp. 2648-2662.

Ahrens, J.P., 1989. Stability of Reef breakwaters. *Journal of Waterway, Port, Coastal and Ocean Engineering, ASCE,* **115(2):** 221-234.

Baba, M., 1985. Design of submerged breakwater for coastal protection and estimation of wave transmission coefficient. Proceedings of 1st National Conference on Dock and Harbour Engineering, Vol 1, I.I.T Bombay, B13-B26.

Cox, J.C. and Clark, G.R., 1992. Design development of tandem breakwater system for Hammond Indiana. Proceedings of Coastal structures and breakwaters, Thomas Telford, London, pp.111-121.

Dattatri, J., Sankar, N.J. and Raman, H., 1978. Performance characteristics of submerged breakwaters. Proceedings of 16th Coastal Engineering Conference. *ASCE,* pp. 2153-2171.

Dick, T.M. and Brebner, A., 1968. Solid and permeable submerged breakwaters. *Journal of Coastal Engineering in Japan,* **72:** 1141-1158.

Gadre, M.R., Poonawala, I.Z. and Kudale, M.D., 1992. Stability of rock armour protection for submarine pipelines, Proceedings of 8th Congress of Asian and Pacific Division of IAHR, CWPRS, India, Vol III, D-149-D-160.

Hunt, I.A., 1959. Design of seawalls and breakwaters. *Journal of Waterway, Port, Coastal and Ocean Engineering, ASCE,* **85(WW 3):** 123-152.

Johnson, J.W., Fuchs, R.A. and Morison, J.B., 1951. The damping action of submerged breakwaters. Trans. *AGU*, **32/5:** 704-718.

Shirlal, K.G. and Rao, S., 2003. Laboratory Studies on the Stability of Tandem Breakwater. *Journal of Hydraulic Engg., Indian Society for Hydraulics*, **9(1):** 36-45.

Shirlal, K.G., Rao, S. and Prasad, S.K.M., 2003. Stability of Tandem Breakwater. *Journal of Institution of Engineers (I)*, CV, **84:** 159-164.

Nizam, Y.N., 1996. Artificial reef as an alternative beach protection. Proceedings of Tenth Congress of Asian and Pacific Division of IAHR, Langkawi, Malaysia, 422-430.

Pilarczyk, K.W. and Zeidler, R.B., 1996. Shore evolution control and offshore breakwaters. A.A. Balkema Publishers, The Netherlands.

Smith, D.W., Sorensen, P.S., Nurse, R.M. and Atherley, L.A., 1996. Submerged crest breakwater design. Advances in Coastal structures and breakwaters. Thomas Telford, London, 208-219.

Twu, S.W., Liu, C.C. and Hsu, W.H., 2001. Wave damping characteristics of deeply submerged brerakwaters. *Journal of Waterway, Port, Coastal and Ocean Engineering, ASCE.* **127(2):** 97-105.

Lake Chilika: GIS and the Challenge of Spatial Management

Geoffery J. Meaden

Canterbury Christ Church University
North Holmes Road, Canterbury, Kent, CT1 1QU, UK
g.j.meaden@cant.ac.uk

1. INTRODUCTION

The use of Geographical Information Systems (GIS) for a range of terrestrial spatio-temporal analyses has grown exponentially over the past three decades. This growth has occurred as a result of the system's proven capabilities at resolving an ever widening range of problems, plus the fact that the system can be utilised for modelling and other predictive purposes. Growth has been supported by rapidly decreasing costs and by the promotion of GIS through publications, education and conferences. The use of GIS for analyses or modelling in aquatic environments has been rather slower to emerge. This is largely a function of the fact that, from a cost viewpoint and from a 'mapping capability' perspective, data here is much more difficult to acquire. Thus, in aquatic environments, everything moves or is mobile including both the environment itself and those objects within it. Despite these difficulties, for the past 15 years GIS has been increasingly utilised for work in both marine and freshwater environments, and there is now a huge literature that can testify to its success[1].

One of the most significant lessons learned from the Chilika experience is the vital role that scientific information can play towards achieving good management. (Ghosh, 2003). Here is recent recognition that the Lake Chilika area is a classic case where innovations such as GIS offer the potential for huge management advantages and progress towards the assumed goals of

---

[1] A paper giving a good overview of the use of GIS in coastal zone management (Zeng, T.Q. et al. (2001) Coastal GIS: Functionality Versus Applications. *J. of Geospatial Engineering*. Vol. 3(2), pp. 109-126) is available on http://www.lsgi.polyu.edu.hk/sTAFF/zl.li/vol_3_2/05_zeng.pdf

sustainability and restoration. The author goes on to note that "the lessons learned (from all the previous studies in the lake catchment) show how scientific research and application can lead to better management in a wetland ecosystem." The issues of how to achieve better management through the realisation of a dedicated GIS for the Chilika lagoon form the substance of this paper. In more detail, the paper will attempt to further elaborate why a GIS is necessary for improved lake and catchment management, and it will exemplify the range of potential tasks that a GIS might accomplish. However, accomplishing these tasks will not necessarily be easy; so it is important that any instigators of a possible "Chilika GIS" are cognisant of some of the challenges that must be faced. The paper concludes with some ideas on a potential implementation strategy for a "Chilika GIS".

2. WHY UTILISE A GIS FOR THE MANAGEMENT OF LAKE CHILIKA?

Initially, it might be useful to mention that virtually all of the problems that are manifest in the catchment area of Lake Chilika originate from the terrestrial areas, i.e. from the land. However, most of the problems are actually manifest in the lake. This gives a vital clue as to the direction in which the attention of managers should be directed. If the land areas, from the perspective of sustainability, can be successfully managed, then most of the aquatic problems will be coincidentally solved. So any "Chilika GIS" must be concerned with the whole catchment of the lagoon.

In the total lagoon (lake) catchment there is a matrix of inter-related problems that lead to major spatial dis-equilibrium[2]. Problems may be manifest (and illustrated) as 'causal linkage chains'. For instance, we can illustrate two of these by tracing backwards to an original cause:

(a) *Sedimentation.* Excessive silting > stream sediment load > inflow via catchment run-off > increased plowing/soil exposure > natural vegetation removal > need for cleared land plus fuel and timber > land needs > population increase.
(b) *Wildlife depletion.* Hunting or over-fishing > need for food > population increase.

There are many other chains, some of which are simple and others are complex. However, they are nearly all interlinked and ultimately derive from population pressures. This is, of course, a large problem that faces both India and the planet!

[2] The inter-related nature of the problems was recognised by the Ramsar Advisory Mission who visited the lake in December, 2001. Their concerns have been noted at the web address: http://ramsar.org/ram/ram_rpt_50e.htm

So, the natural ecosystems in and around the lake are out of balance. What makes the Chilika case so important is the scale and the range of the problems in terms of numbers of ecosystems affected, the numbers of humans affected and the degree to which the systems are out of balance. To alleviate the situation here is a truly major challenge but it is a challenge that must be taken on. Of course, there has been much recognition of these problems, and especially during the past decade some considerable amount of research and management work has been directed towards improving conditions in the catchment area (Sudarshana, 1999). However, the aforementioned dis-equilibrium, in conjunction with the multiplicity of the 'causal linkage chains', means that quite sophisticated remedial measures must now be deployed. This is where GIS comes in. As an analytical and decision-making tool, GIS can provide managers with the means to rectify situations where spatial imbalance prevails. Because it handles data relative to both spatial and temporal dimensions, and through its multiplicity of analytical functions, a GIS can be at the heart of any decision-making forum.

Management decisions on the future of the lake cannot be taken without access to a very wide range of data and information. There are already a number of mechanisms in place to gather such data. However, at present it appears that there is an over-reliance on data from satellite remote sensing (RS). Whilst RS data is of course extremely useful, it can only provide information on a restricted range of parameters (e.g. Pal and Mohanty (2002) report on aquatic vegetation cover, and Mohapatra et al. (1994) report on other variations in lagoon features). For its successful functionality, GIS is totally dependent on accurate spatially referenced data, and for managers to successfully utilise GIS across a wide ranging number of thematic areas, then a wider range of data gathering systems must be put in place. Instigation of a GIS for the improved management of the Chilika catchment would act as a significant spur towards achieving the emplacement of further objective data gathering systems.

The enactment of improved data collection, in conjunction with the GIS itself, can act as a significant unifying force for good. Thus, there has already been put in place a group of trained facilitators who are locally running some 10 Centres for Environmental Awareness and Education (Ghosh, 2003). This was a response to the perceived need to make the local population aware of the pressures on the lake environs. These centres, plus the Campaign for the Conservation of Chilka Lagoon, might need to further organise local ground-based data collection. It is likely that most of those who come into contact with these facilitators will unite in their efforts to bring about improvements. It is likely that additional, cooperative data gathering mechanisms will be adopted. Thus it makes financial and strategic sense to utilise some of the thousands of 'feet on the ground' as a means of data collection, especially where it involves collection of data relative to human-based activities. In many cases groups who might have previously been actively hostile to one another, or who would certainly not have actively cooperated, might in the future be interacting.

There are some indications that much of the scientific work previously carried out on the lake and its environs is descriptive or one-dimensional (Mishra et al., 2003). For instance, there are large data sets that elaborate on the number of *x* or *y* species found in the lake or the catchment. To an extent the fourth dimension has been introduced via some examination of species and their quantitative changes over time. Although Pal and Mohanty (2002) correctly recognise that process analyses (rather than simply descriptive) are increasingly desirable, it is imperative that studies are taken further through the adoption of both two and three dimensional analyses, i.e. in association with temporal considerations. GIS offers this opportunity.

A further reason for the implementation of a GIS for the management of the Chilika catchment lies in value of mapping *per se*. In the area of cognitive perception there is some considerable evidence that a 'picture is worth a thousand words'. The same cognitive processes underlie spatial comprehension. Whilst of course there will be significant inter-personal differences in spatial (mapping) cognition, it is very clear that a map has the potential to instantly convey a range of spatial relationships and process events. This might best be exemplified by a weather forecast map. Here the user has little difficulty in perceiving meteorological processes (sunny, raining, hot, windy, etc), as well as where these process events might be in relation to known geographic locations. Extrapolating to the Lake Chilika circumstances, it is easy to see that processes or activities in the region can also be visually conveyed through the media of mapping.

Another justification for a "Chilika GIS" lies in the system's inherent capability of being scale independent. Thus, all the process interactions inferred above will be happening across a wide range of scales. This means that there is no optimum scale at which a "Chilika GIS" might operate, and indeed it is essential that it does function at a wide range of scales. So, for some variables (parameters) it might be best to carry out investigations into specific small areas, but for others this will be unnecessary. GIS is able to change scale if required and different data sets from different areas, or data showing different processes, can be integrated into any desired analyses.

There are a range of other factors that also favour adoption of a GIS for the management of the catchment. These include the rapidly growing capacity for India, through the use of locally trained people, to provide the technical backup for the system; the ability to secure a range of data that easily integrates to GIS in terms of formatting and structure; and the fact that a successful adoption here would be a major incentive to GIS adoption for environmental management in other parts of the sub-continent that are experiencing spatially related conflicts. GIS has been most successfully used in similar situations elsewhere[3], so there is the strong possibility that a "Chilka GIS" would be able to benefit from the experience of others. And finally, the Ramsar Convention, in its new guidelines for the integrated management of wetland planning, notes that a major process should be "to begin development of (a) GIS database and other area management tools"[4].

3. POTENTIAL GIS USES IN THE CHILIKA CATCHMENT AND DATA REQUIREMENTS

For those who have been working with GIS for any length of time it is clear that, in any specific field for spatial analyses, there may be an almost infinite range of analyses that could be performed. The situation at Lake Chilika is no exception, and indeed the uses to which a GIS could usefully be applied here would be far greater than in most other circumstances. This is simply because the lake and its environments are encapsulating a microcosm of almost all human activities. Given this wide potential, an exemplary list can only briefly illustrate potential GIS uses:

- As a means for various kinds of mapping. This might be of static features such as boundaries or roads, or mobile features such as fish distributions or lake chlorophyll, or features exhibiting irregular patterns of change such as forest clearance or land use.
- To examine the relationships among spatial distributions of phenomena associated with the lake. The GIS can reveal what variables are related, the strength of the relationship(s), and how the relationship(s) may vary over time.
- Resource distributions need to be known in order to identify potential extraction quantities, e.g. fish stock assessments will capture data that can help ascertain spatial variations in fish exploitation, and where exploitation might best be located.
- It will be essential to establish both reference points and indicators for the monitoring of the relative success of management plans. Basically, this is the setting of a baseline(s) against which it is possible to judge whether progress in lake management is being achieved. Indicators might be descriptive, performance, welfare or efficiency based. It is important to note that, given the increasing human populations around the lake, it will be impossible to re-attain some idealised, pristine conditions for the lake. GIS can give an objective indication on the extent to which performance targets have been met.
- There is now a growth in developing habitat suitability indices based upon known environmental preferences for given species (see Fig. 1).

[3] An example of a very similar project can be found at: http://www.geog.ucl.ac.uk/melmarina/work_packages.stm

[4] For further details see: http://www.ramsar.org/strp/key_strp_workplan_2003.htm

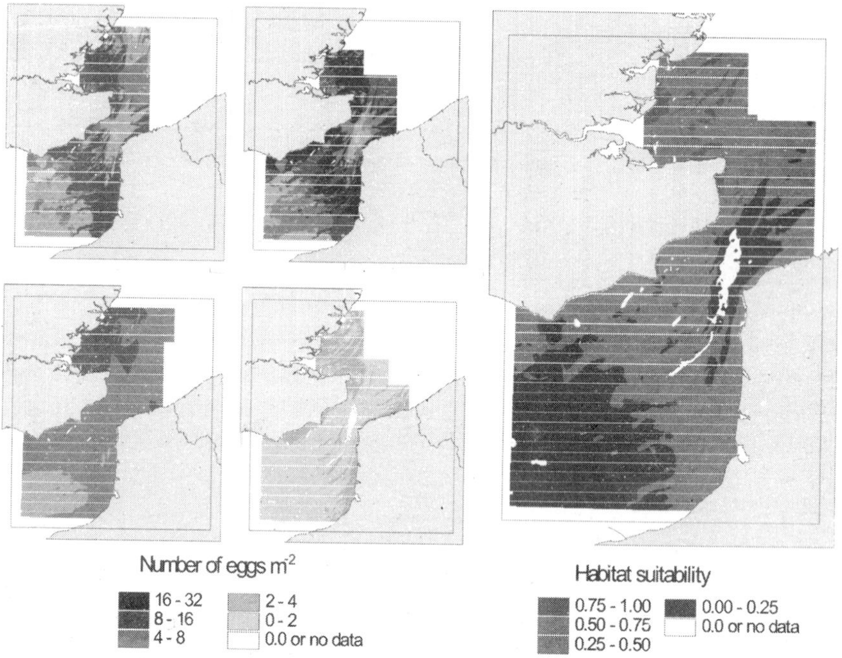

Figure 1: Developing a habitat suitability model for sole eggs in the Dover Straits (adapted from Eastwood et al., 2003).

This is essential for the establishment of, for instance, 'no-take zones' or protected areas for possible fishery closures.

- GIS can usefully monitor the effectiveness of closed areas both spatially and temporally, also giving clues to the desirable size, shape and disposition of such areas.
- GIS can aid in the identification of areas that are in need of remedial measurements such as habitat restoration, e.g. areas where silt is accumulating or where vegetation growth is a problem, and areas where mangroves could advantageously be replanted.
- It will be possible to develop process models. GIS can form the working platform for these, and there are a range of external models that can be linked to most proprietary GIS's[5]. Using dynamic (animation) GIS techniques, it is possible to display models of processes such as seasonal water dynamics, sediment transport, salinity variations, water volumetric calculations, vegetation dynamics, and various run-off trajectories[6].

---

[5] The paper accessed on http://www.coastgis.org/01pdfs/gilman.pdf gives valuable clues to integrated modelling.

[6] There are a number of web sites that are able to supply environmental software models that in many instances can be integrated to function in a GIS environment, e.g. http://www.spatialhydrology.com/software_hydrostat.html and http://www.uvm.edu/giee/SME3/

- It will be possible to test various restocking strategies including measuring the success rate of these and establishing where restocking needs are most needed.
- It is desirable to estimate optimum location for such physical structures as processing plants or prawn farms (if this activity is permitted in the lake), or for activities such as eco-tourism. Use of GIS for these studies might be thought of as optimal location or impact analyses.

It is clear that many of these tasks will be essential to the future enhancement prospects of Lake Chilika. Table 1 shows the 'scientific needs' for the successful integrated management of Chilika as suggested by Mishra et al. (2003). It is also clear that GIS can significantly contribute to the holistic attainment of these needs, i.e. since nearly all of them have a spatio-temporal context. What is further clear is that, for the success of most of these analyses, it will be necessary to significantly enhance the range of data that is presently held.

Table 1: Scientific needs for the sustainable management of Lake Chilika (from Mishra et al., 2003)

1.	Regular monitoring of the lake
2.	Better techniques for aquaculture
3.	Fish recruitment and assessment of stock
4.	Fertiliser and sewage related pollution
5.	Heavy metal pollution
6.	Conservation of avifauna
7.	Dredging affect on species diversity
8.	Research and database

There is already assembled a quantity of RS satellite data and this can provide excellent information on factors such as lake morphology, distributions of water-borne vegetation, lake-side land use, water turbidity and seasonal changes in water quantity, etc. By dint of its collection methods, this data is accurately geo-referenced. However, a problem regarding much of the rest of the data currently being held is the quality of its geo-referencing and this is something that future data gatherers must be aware of. From a rudimentary perusal of the Chilika situation it appears that the data that are presently lacking are mainly those associated with human activities, e.g. the distribution of fishing effort, applications of chemicals or fertilizers to farming (plus other inputs to the land), eco-tourism dispersion, plus those associated with marine and terrestrial faunal distributions.

Although this author has no way of knowing the exact data sets that are currently held, examples of data that could form the basis of much GIS work include:

- A time series for lake shore dynamics (both seasonally and over the long term).

- Lake bathymetry.
- Long term and seasonal distributions of water qualitative abiotic factors such as, pH, salinity, temperature, nutrients, etc, plus distributions of biotic factors, e.g. chlorophyll, BOD, planktons, vegetation.
- Species or habitat distributions - marine and terrestrial.
- Catchment land use including temporal changes.
- Catchment population density and changes.
- Boat concentrations, distributions and density.
- Catch statistics and distributions.
- Facilities locations, e.g. prawn farms, processing plants, road access, jetties, supply points, etc.
- Transport systems.
- Location of infrastructure.
- Point sources of pollution.
- Restricted zones.

It is important to stress that these are only some of the major thematic areas for data sets—within any of these categories numerous sub-categories could be defined.

4. CHALLENGES TO THE USE OF GIS IN A DYNAMIC ENVIRONMENT

Challenges to the use of GIS for work in areas such as the Chilika basin are going to be widespread, i.e. if only because of the all-embracing nature of the human and physical environments involved. However, recognition of these is imperative if the GIS is to be successfully deployed. This recognition is necessary not only as an aid to the progression of successful GIS output, but also in order to best recognise the limitations that may be placed upon any of the interpretation of output from any specific GIS projects. Challenges can best be looked at under the four main headings of: Intellectual and theoretical; Practical and organisational; Economic; and Social and cultural.

4.1 Intellectual and Theoretical Challenges

4.1.1 The Mapping of Moveable Variables

The first intellectual or theoretical challenge is that of mapping moveable variables. Unlike terrestrial environments, in a marine area both the environment and everything in it moves. Only on the sea floor are there a number of unmoving reference points. Some water movements can be well predicted and measured, and these can be modelled in a GIS (in terms of speed, volume and direction), but other movements are chaotic and therefore difficult to model and incorporate. In the Chilika situation the problem is exacerbated because lagoons are exceptionally dynamic physical features whose life expectancies may be >1000 years. In Chilika there has been recent infill caused by rapid sedimentation processes, but rising sea levels may result in future breaches of the coastal bar with consequent destruction of this feature and the eventual disappearance of the lagoon. Similar process and movement problems exist with mapping animal (fish or other marine fauna) movements.

4.1.2 Multiple Scale and Resolution

Within the marine environment, processes operate (or objects occur) at greatly varying scales. This means that each GIS project must carefully evaluate what scale it should work at. The problem of resolution of work is then introduced, i.e. measuring the spatial scale and temporal period at which the data has been (or should be) gathered. The challenge here is to identify relevant scales and resolution, and then to develop an ability within the GIS to function at a range of scales. This can be a difficult challenge, but one that is vital for the success of any GIS work.

4.1.3 Handling Three and Four Dimensions

This challenge is best envisaged diagrammatically. Figure 2 shows a hypothetical 3D volumetric domain. Time is the fourth dimension. The block diagram conceptualises the challenges associated with working in more than two dimensions. Here the marine area is divided into cuboids that vary in

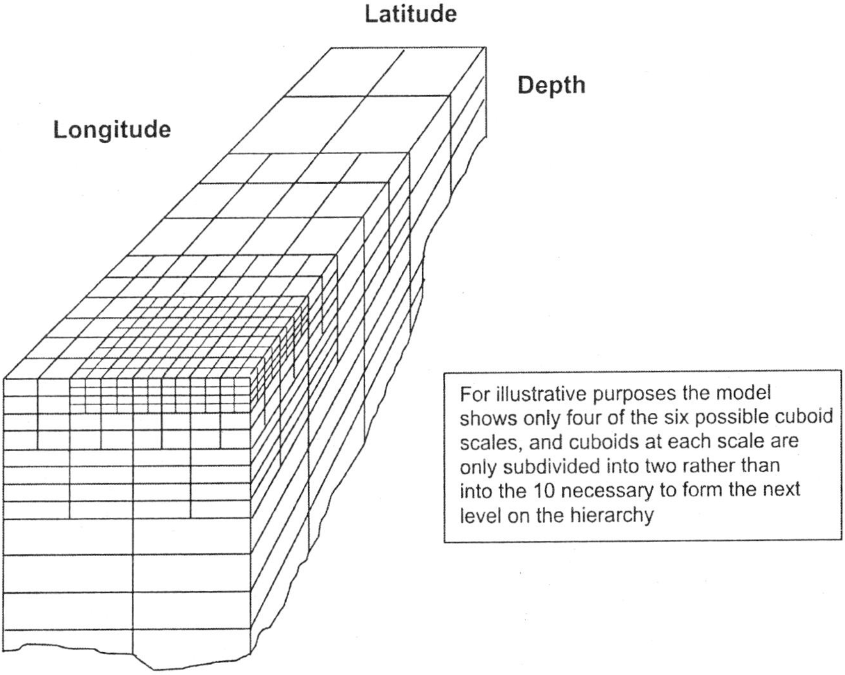

Latitude
Depth
Longitude
For illustrative purposes the model shows only four of the six possible cuboid scales, and cuboids at each scale are only subdivided into two rather than into the 10 necessary to form the next level on the hierarchy

Figure 2: A hypothetical 3D hierarchical structure for data storage.

size from 10,000 × 10,000 × 1,000 metres to small cuboids measuring 0.1 × 0.1 × 0.01 metres. Anyone working in the marine environment has to decide which is the appropriate resolution to be working at? The problem might be considered of going out into the lake and lowering, for instance, a thermometer into the water to take a temperature reading. Where should the next reading be taken? Is the reading valid at a range of scales? Can a single data gathering strategy populate the database in all dimensions at each resolution? It is important to establish the relationship between the sampling frequency and the natural variation in an attribute (Lucas, 2000). A large number of terrestrial databases (and GIS's) can handle three dimensions, but there are very few marine database structures that can generate GIS output that can cope with the complex visualisation involved.

4.1.4 Application of Spatial Statistics and Models

Ecologic or environmental systems are subject areas that have attracted an abundance of mathematical models and applications of spatial or temporal statistics; these are usually explanatory or predictive models. GIS software is well suited to providing an activity surface for modelling. For instance, you can imagine that it is easy to develop an equation that plots (models) the rate (speed/number/direction) of a marine organism as it moves through a salinity or temperature gradient. However, presently only relatively simple models can be adopted by GIS, and it is a major challenge to incorporate some of the more complex fisheries models or sedimentation simulations, e.g. stock assessment, environmental or population models. The important factors of spatial autocorrelation (establishing the relationships between multiple variables), and that of overcoming the lack of statistical significance in most marine data, are likely to be very severe challenges for the future. However, it is likely that environmental modelling will be made easier in the future by the linking of external specialist software to GIS so that the processing is done outside of the GIS.

4.1.5 Optimising of Visualisation

This is the final intellectual challenge. It is essential that output from a GIS is visually meaningful to a range of users and decision makers. To achieve this, the main visualisation challenges are concerned with:

- Classification in terms of a numerical range.
- Classification in terms of categories of any parameter or variable.
- Aesthetics of colour, fonts, symbology, etc.
- Map clarity, e.g. placement of labels and lettering.
- Fuzzy boundary definitions.
- Periodicity of mapping any movements.
- How to portray complex marine surfaces in map form, e.g. coral reefs or mobile and patchy distributions.

All of these visualisation challenges can be tackled but there are never likely to be definitive answers simply because each individual has his own visual perceptual preferences. Figure 3 shows that it may be difficult to show the best way of displaying a map.

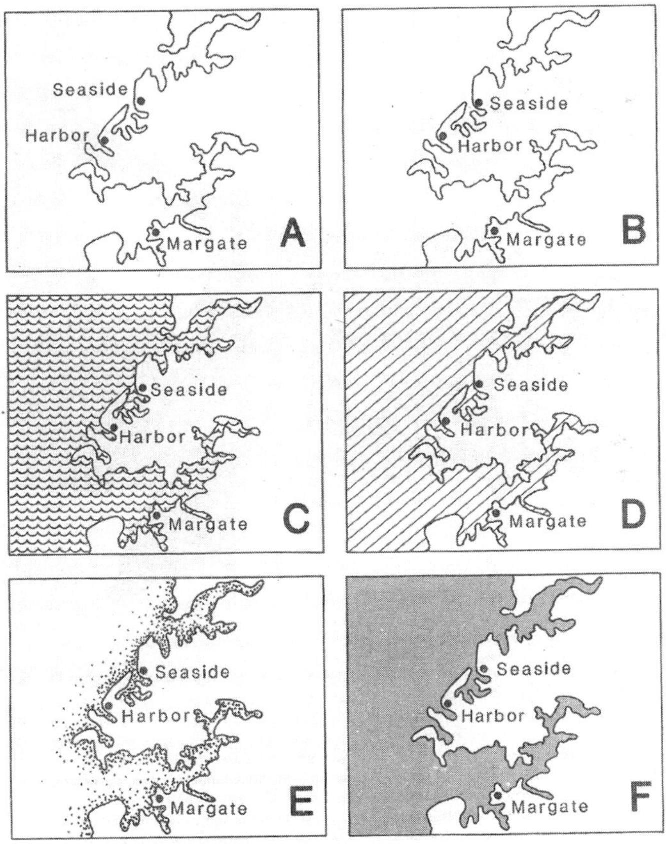

Figure 3: Six different visualisations of a simple lagoon side setting.

4.2 Practical Organisational Challenges

4.2.1 Data Gathering

Data assembling is a challenge to all GIS work, but it is especially difficult for marine and/or environmental work. It is difficult for all the reasons outlined above (space/time resolution considerations, movement, three and four dimensions), but also because of the uncertain data needs, the high costs of data collection, perhaps charter vessel availability, weather problems, time problems, sampling strategies, etc. So, a huge effort is needed to obtain what may be a comparatively limited range of primary data. And when the data has been collected, how statistically significant is it? How replicable would the data be if further counts were made? Are all collectors using the same methods and standards? Although lots of secondary data is supposed to be available on the Internet, this author has yet to download a single piece of useable data! It seems important that in the Chilika situation you are able to utilise the large reservoir of human resources, especially since many of them will have an intricate knowledge of the lake environs.

4.2.2 Subject Organisation

This challenge is concerned with the ways in which the subject field of coastal and/or marine science is organised. The subject is very diverse and is concerned with environmental and fishery science and management, biology, meteorology, geomorphology, oceanography, economics and various technologies. Research into the subject may be pursued by a huge range of often small and dispersed institutions. So, the whole subject area is very fragmented. This means that there may be many small projects going on, with relatively low budgets, with little cohesiveness in project aims, and there is undoubtedly a great deal of duplication of effort. And the dissemination of information from various research projects itself is often fragmented and confined to the grey literature

4.3 Economic Challenges

Here the concern is with the costs associated with marine or coastal GIS and with obtaining the funding to cover these costs. Although the costs of hardware and software for implementing any GIS have greatly decreased, data costs have risen significantly both in real terms and as a proportion of total costs. Data acquisition now consumes an average of 80% of costs for any GIS project, and it is likely that this proportion will rise in the future. In developing countries such as India, the economic situation may be much worse, and a GIS may be an expensive luxury, typically only possible through donor support. Funding may be hard to obtain because it is difficult to demonstrate cost:benefit advantages for a GIS, i.e. how can it be proven that sophisticated GIS-based output can substantially enhance the decision making process?

4.4 Social and Cultural Challenges

4.4.1 Overcoming Technologic Inertia

Overcoming inertia to the implementation of new technologies is certainly understandable, and it might relate to the cultural ambience of the group, company, organisation or region. So, efforts to implement GIS may be difficult because the system may have no one who is available to promote its use. Questions may be asked such as "why should we pay lots of money simply to acquire this complex computer technology that only produces maps?" Also, many fisheries or environmental personnel do not appreciate the fact that their research and management problems are based upon spatial dis-equilibrium, and in some cases they might not know of the existence of GIS.

4.4.2 Achieving Cultural and Directional Cohesion

In the prevailing human milieu of the Chilika catchment area, there are a very wide variety of cultural, religious, social, national and educational classes and groups. Each of these groups may have their own life styles, ambitions, and aspirations, and it is known from past experience that these are not always in sympathy. Ghosh (2003) recognises many socio-economic challenges facing lake management such as weak government institutions, the fragmented approaches to management, local corruption and the lack of effective enforcement strategies. But, there is only one Lake Chilika, and if it is to continue to sustain a large and growing population at higher living standards, then there has to be consensus on its preservation, and indeed on its restoration. This will entail the simple concept of 'working in harmony'. From the GIS perspective this means that management decisions for the lake will need to be agreed, i.e. so that GIS projects, with their data gathering requisites, can be formulated. It will be essential that all users of the lake be participants in future decision making processes.

4.4.3 Developing Geographic Cognition

Finally, but of major significance, is the challenge to successful GIS implementation caused by a lack of geographical cognition. This challenge is concerned with having an appreciation of the geographic aspects of the problem, plus an understanding of geographic relationships, and the ability to readily comprehend geographic inputs, analyses and outputs. As the author has said elsewhere "Geographic expertise involves the recognition of spatial relationships in terms of adjacency, ubiquity, heterogeneity, contiguity, etc. Visual discrimination is a vital tool (or ability) in seeking out these subtle relationships. Likely flows and interactions must be recognised and understood, as must spatial patterns, surface trends and zonal forms" (Meaden, 2001). Figure 4 shows a world map as presented by a world fisheries authority, illustrating that they were probably not cognisant of 'real geographies'! There is a huge challenge to imparting this appreciation, probably because it is largely an innate skill.

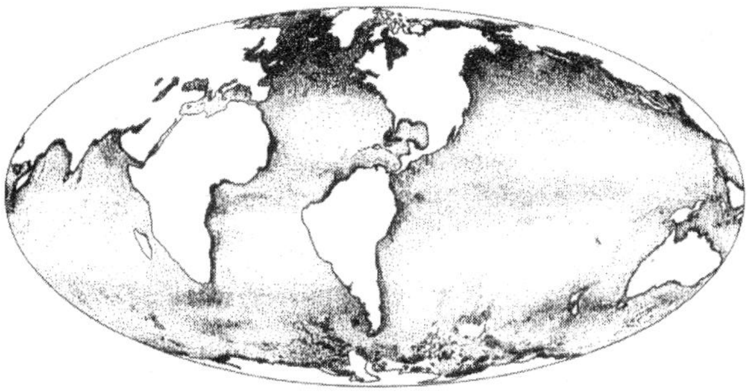

Figure 4: The world, as depicted by leading fisheries authorities, showing lack of geographic cognisance (Caddy and Griffiths, 1995).

5. DEPLOYING A GIS TO MANAGE THE CHILIKA LAKE

Although the author is not in a position to know the full extent of work that has been done towards implementing a "Chilika GIS", it is apparent that complementary, peripheral work has been accomplished. He is also not in a position to know the extent of any funding that might or might not be available, but the following scenario for GIS implementation will be couched in terms of what might be realistically achievable. In this section an overview is given of the strategy that needs to be undertaken in order to achieve successful GIS adoption. In making these suggestions there is an implicit assumption that a dedicated GIS is seen as desirable, that overall system's goals are identified, that management support is strong, and that all the following major steps are taken under the guidance of a proven expert. The steps are set out in a logical order of priority, but this will not prevent work on several of the implementation strategies being followed together:

1. *Needs recognition:* An essential early task would be the setting up of a range of realistic aims and objectives for a "Chilika GIS". These might be couched in both general terms or in very specific projects. What are the urgent spatially related problems? This task will clearly revolve around a large number of experts, but it should be careful to include inputs from local lake users and lakeside dwellers. Indeed it might be necessary to indulge in further 'catchment coalition building' to ensure that the needs of all interested parties are met or are at least considered. At the conclusion of this task it might be necessary to undertake a cost:benefit analysis, i.e. to see whether or not to still take the GIS forward.
2. *Organisation of the system:* If a GIS is seen to be desirable then a dedicated 'Chilika GIS Centre' might be established at a location perhaps chosen by the Chilika Development Authority (CDA). This facility could take various formats in the sense that it could be sited at one location as a stand alone system, or the GIS could be distributed such that various locations of excellence were interlinked enabling expertise to be best utilised. For example, RS expertise could be utilised at the University; perhaps ecosystems datasets were collected and stored at a conservation centre; and a hydrological centre might manage water parameters. Each of these centres would be responsible for the GIS work, and external connections to a variety of users and contributors would then need to be functionalised.
3. *Appointing a GIS 'champion':* Fundamental to the success of any GIS project is that it is directed by an individual who is completely enthusiastic, dedicated and competent. This person must not only be broadly conversant with the technical demands of GIS but he or she must (in this case) be familiar with the Chilika situation and be sympathetic to the goals of its managers and their strategies.

4. *A data repository:* Data collection, management and upkeep would be an essential basic task. Decisions on whether this data should be deposited in a central repository, or whether the individual data collectors (or institutions) should remain responsible, must be made. These decisions could have a great effect on how the whole system physically functioned and what inter-institution connectivity was required.
5. *Acquiring the system and operatives:* Dedicated staff need to be acquired and possibly trained, and dedicated hardware and software needs to be agreed upon and obtained. This would be an ongoing task, but one of major importance in a subject area that is undergoing rapid evolution. Although it would be possible for the development of a specialised "Lake Chilika GIS", i.e. a software package explicitly designed for tasks perceived as important to the Chilika situation, the author would very much caution against this. This is because it is sensible to invest in an 'off-the-shelf' GIS product initially, certainly until such time as major priorities had been decided. Also, this would be a much cheaper strategy.
6. *Setting up a web-site:* From the start a 'Chilika GIS' web-site needs to be established. The importance of this lies in publicising what the system is all about. It is also vital to get a range of people on your side. This task brings with it the necessity of constant web site updating.
7. *Prioritising GIS tasks or projects*: It is essential to commence the GIS work with fairly simple tasks. This is to do with both the learning process and the importance of first undertaking work that might require relatively little data. It is also important to maintain staff morale, and successful outcomes based on more simple tasks can best achieve this. Getting early successes is also important with regard to publicising the work that is done, both to other associated workers and to the various parties who have an interest in improving overall conditions for Lake Chilika.
8. *Scale of the study:* It will be important to establish an optimum scale of study for each parameter. This is because some parameters vary markedly over short distances whereas others are comparatively uniform. In combination with this there will be a consideration of the geographic extent of individual parameters or data sets.
9. *Data gathering strategies:* Once project needs have been identified, then the data needed to fulfil the project can be listed. Data gathering strategies then need to be optimised, and there are several ways of achieving this. For some GIS tasks it might be preferable to start from a 'whole lake' perspective using just a few important parameters. But, for other GIS tasks, a decision might be made to approach data gathering for a number of small, possibly problem thematic or geographic areas. Ghosh (2003) indicates that there are already 20 'micro watershed management' areas within the lagoon catchment, and these could usefully serve as sub-data regions. There are advantages and disadvantages for

each approach. Considerations need to be made on how the data was to be acquired or gathered, and this would be very much dependent on the availability of funds. The options here are vast and space forbids a further discussion of this.

10. *Working with others:* Having pointed the need for a 'basic beginning', it is quite likely that a "Clilika GIS" will wish to work in collaboration with a local centre of excellence (often a University), and they should be encouraged to participate in joint projects. This collaboration will help both partners and it may spread both the funding and expertise burden. Here some more sophisticated tasks, such as process modelling, might be attempted.
11. *External funding:* Given the scale of the problems that Lake Chilika faces, and the sheer size of the Lake plus its multi-faceted problems, the author feels that funding and help for a Chilika GIS should be requested from the FAO in Rome (via perhaps the state government). Failing this, other alternative international agencies should be approached. There is, doubtless, already sufficient state or local expertise in this funding acquisition field.

6. CONCLUSIONS

The case for a "Chilika GIS" is overwhelming, and the catchment area has all the ingredients for its use. There is the recognition that the area has a tremendous range of spatially related problems and that the only way to overcome these is through consensual management. Good management is based upon being able to make decisions in the light of a range of appropriate information, and there are already a number of programmes and research studies in place aimed at drawing attention to the plight of Chilika. GIS is a proven technology that can be expeditiously utilised for the provision of requisite information. In India there are many bright, young and perceptive computer workers who could be engaged in the task of operating and managing a dedicated GIS. And the problems for the lake are relatively concentrated, making the task of data collection easier. All of these factors combine to indicate that GIS should be adopted as part of a lake management policy designed to prolong and sustain the natural and human environs of the lake.

In his recent study of Chilika, Ghosh (2003) commented that "the use of scientific databases, RS` technologies, GPS mapping, mathematical modelling and the continuous monitoring for hydro-biological parameters and for the impact of dredging on the lake's ecosystems will provide tangible benefits in the management process." GIS is missing from this list! Why not provide the very platform that links all of these advances together?

REFERENCES

Caddy, J.F. and Griffiths, R.C., 1995. Living Marine Resources: Their Sustainable Development. *FAO Fisheries Technical Paper No. 353*. FAO, Rome.

Eastwood, P.D., Meaden, G.J., Carpentier, A. and Rogers, S.I., 2003. Estimating limits to the spatial extent and suitability of sole (*Solea solea*) nursery grounds in the Dover Strait. *Journal of Sea Research*. **50(1/2):** 151-165.

Ghosh, A., 2003. Experience brief for Chilika Lake, Orissa, India. Centre for Environment and Development.

Lucas, M., 2000. Representation of variability in marine environmental data. *In:* Wright, D.J. and Bartlett, D.J. (Eds) Marine and Coastal geographic Information Systems. Taylor and Francis, London. pp. 53-74.

Meaden, G.J., 2001. GIS in fisheries science: Foundations for a new millenium. *In:* Proceedings of the First International Symposium on GIS in Fishery Science. Nishida, T, Kailola, P.J. and Hollingworth, C.E. (Eds). Seattle, Washington, USA. March 2-4, 1999. pp. 3-29.

Mishra, P., Mohanty, P.K. and Sugimoto, T., 2003. Environmental conditions and strategies for sustainable management of Chilika Lake, India. *In:* Asia and Pacific Coasts, pp 14.

Mohapatra, K.N., Sudarshana, R. and Das, N.C., 1994. Remote sensing based study on spatio-temporal variabilities of the coastal lagoonal features: A case study of Chilka Lake, India. Proceedings of the Coastal Zone Canada '94. Halifax, Nova Scotia, Canada. vol.3; pp. 936-949.

Pal, S.R. and Mohanty, P.K., 2002. Use of IRS-1B data for change detection in water quality and vegetation of Chilka lagoon, east coast of India. *International Journal of Remote Sensing*. **23(6):** 1027-1042.

Sudarshana, R., 1999. Multiple use of a coastal lagoon: Success and failure/Chilka Lagoon - India. (accessed 28.11.2004 on www.csiwisepractices.org?read=38).

Mapping Lagoonal Features and Their Variability: Field Observations and Remote Sensing Implications

U.S. Panda, P.K. Mohanty, S.R. Pal, G.N. Mohapatra, P. Mishra[1] and G. Jayaraman[2]

Department of Marine Sciences, Berhampur University, Berhampur 760 007
mail2uma@rediffmail.com
[1]ICMAM Project Directorate, Department of Ocean Development, NIOT Campus, Velachery-Tambaram Highway, Pallikaranai, Chennai-601 302
[2]Centre for Atmospheric Sciences, Indian Institute of Technology New Delhi-110016

1. INTRODUCTION

Lagoons have been historically important as sheltered sites of habitation providing access to both the land and the sea, comprising 15% of the world coastal zone, in which lakes, salt marshes, tropical mangroves, swamps and deltas are also included. Natural changes resulting from physical, chemical, geological, and biological factors and the influence of climatic changes will alter the essential character of the lagoon and hence the ecosystem. For example, excessive runoff due to flood during heavy rainfall may cause a chain of events like increase in sediment load leading to turbidity which tends to reduce sunlight penetration which in turn gives rise to lowered primary productivity. Long term unpredictable climatic variations such as tidal waves, hurricanes, cyclonic storms might also cause irreversible changes through sediment loading, alteration of flushing rates of lagoons and production processes (Fisher et al., 1972). So it is important that the maximum benefit from these areas be obtained without jeopardizing the future options or continued use.

Chilika lagoon—the largest brackish water tropical lagoon in Asia, satisfying the definition of lagoon (Phleger, 1981) and process of formation (Nichols and Alien, 1981)—has been the concern of the scientific community of the region, planners, administrators and politicians. It is because, by nature most lagoons do not persist for long periods in the geological time scale and eventually become swamps, marshes and ultimately disappear by plant colonisation and encroachment (Barnes, 1980).

Due to its vast potential wealth of living and non-living resources and rich biodiversity, Chilika lagoon is considered as a '*Ramsar site*', a status accorded to it by the International Convention of Environment held in Ramsar in Iran in 1971. The Convention on Wetlands came into force for India on 1 February 1982.

The lagoon environment over the past few decades has undergone some visible changes owing to many natural phenomena and anthropogenic pressures. The lagoon was under constant ecological threat due to siltation and sedimentation through river runoff and land runoff from the catchment areas, leading to choking of the inlet mouth and poor interaction with the sea. Thus, siltation, choking of the out channel, shifting of the inlet mouth northwards, decrease in salinity, eutrophication followed by prolific growth of weeds, decrease in fish productivity, shrinkage of water area, loss of biodiversity and increase of human interference through prawn and aquaculture were identified as the major problems with the contemporary phase of lagoon transformation. Owing to the ecological threats that Chilika was facing, the lagoon was added to the *Montreux Record* on 16 June 1993. Montreaux record lists wetlands of international importance, which are already in danger due to environmental degradation. To address the various problems concerning the Chilika lagoon, the Government of Orissa through Chilika Development Authority (CDA), monitored various multi-disciplinary and multi-dimensional developmental activities with an overall objective to protect the lagoon's eco-system with all its genetic diversity and to restore its past glory. Apart from several measures undertaken, one of the major accomplishments achieved by the CDA in this direction has been artificial opening of a mouth of the lake into the sea on 23 September 2000. The lagoon, after opening of its new mouth, witnessed rapid ecological changes (Chilika, 2001) which paved way for its restoration. Chilika was removed from the Montreux record by the Ramsar bureau with effect from 11 November, 2003 for the improvement of the ecosystem of the lagoon after the restoration measures were successfully implemented by the CDA.

Remote sensing offers promise in the detection and delineation of the functional elements of lagoon transformation. Therefore time series Indian Remote Sensing Satellite Data (IRS 1A LISS-I and IRS 1D LISS-III) along with field observations in the post mouth opening period (2001-2004) have been used in the present study for a qualitative and quantitative evaluation of the components involved in the transformation and to obtain estimates of their distribution, abundance and physical state. Pal and Mohanty (2002) in their study pertaining to the pre mouth opening period noted that the seasonal mode of variation in Chilika is dominant over the interannual mode. Therefore, it would be interesting to look into the modes of variability both in the seasonal and interannual mode in the post mouth opening period and to suggest the role of different processes which are in operation in the post mouth era of lagoon environment.

2. DESCRIPTION OF STUDY AREA

The Chilika lagoon (N 19° 28′-19° 54′; E 85° 06′-85° 35′) on the east coast of India in the Orissa State (Fig. 1) is the largest lagoonal system in the subcontinent and is one of the largest tropical lagoons of the world. The lagoon is 71 km long and 3-32 km wide. The water area of the lagoon is variable from a maximum of 992 km^2 during rainy season to 815 km^2 during summer (Mohanty et al., 2001). At the northern end, tributaries of the Mahanadi river, such as Daya, Nuna and Bhargavi join the lagoon (Fig. 1) and are responsible for the large fresh water and sediment influx to the lagoon. The lagoon is separated from the Bay of Bengal by sand bar of 60 km in length. The lagoon has two mouths, a 24 km long narrow and curved channel which runs parallel to the coast to join the Bay of Bengal near Arakhakuda, and the new mouth at Sipakuda at a distance of 8 km from the main body of the lagoon. Table 1 gives the recent physiography and physico-chemical characteristics of Chilika lagoon.

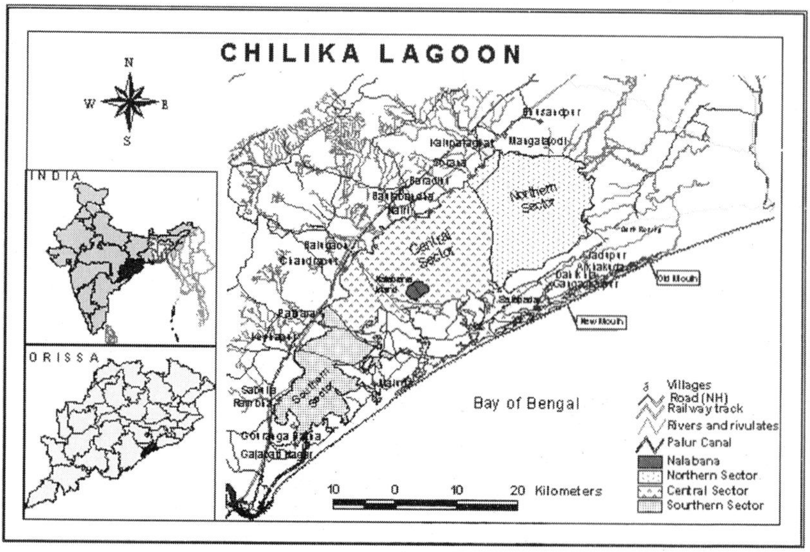

Figure 1: Map of Chilika Lagoon showing different sectors and positions of New and Old mouths.

Table 1: Physiography and physico-chemical characteristics of Chilika lagoon

Physiography	
Location	Latitude 19° 28'-19° 54' Longitude 85° 06' - 85° 35'
Boundaries	East: Bay of Bengal West: Rocky Hills of Eastern Ghats North: Alluvial plain of Mahanadi delta South: Rocky Hills of Eastern Ghats
Water spread area (sq. km)	Summer months: 906 Rainy months: 1165
Length	65 km (in NE-SW direction)
Breadth	16 km (Broadest)
Total catchments	3560 km^2
Shape	Pear shaped
No. of rivers and rivulets draining into the lake	10
No. of islands inside the lake	106
Age	~ 5000 years
Lake mouth	Old mouth (Arakhakuda); 24 km from Satapada New mouth (Sipakuda); 7.3 km from Satapada
Major ecological divisions	Four (Northern Sector, Central Sector, Southern Sector and Outer channel area)
Depth (m)	0.38-4.2—Average: 1.35
Physicochemical properties	
Temperature (°C)	
Surface water	17.5-32.0, Vertical gradient: ~1
Salinity (ppt)	Traces: 36.0, vertical gradient: ~2
pH	7.6-10.0, vertical gradient: <1
Dissolved Oxygen (mg/l)	1.3-13.4
Nutrients (mg/l)	
Nitrate	Traces-0.19
Nitrites	Traces-1.3
Phosphate	Traces-0.18
Silicate	0.10-0.60
Major Elements (ppm)	
Calcium	24-330
Magnesium	87-1380
Sodium	295-13500
Potassium	17-428
Chloride	604-19650
Sulphate	104-3000
Bicarbonate	90-159
Trace Elements (ppm)	
Copper	0.02-0.04
Zinc	0.025-0.19
Iron (%)	0.12-0.32

Source: Panigrahi, 2005

3. METHODOLOGY

3.1 Field Data Collection and Analysis

Observations on water quality parameters such as depth, secchi depth, water temperature, salinity, pH, dissolved oxygen (DO), biological oxygen demand (BOD), water nutrients (ammonia, nitrate, nitrite and phosphate) and chlorophyll were carried out during summer (MAM) and winter (NDJF) seasons covering about 29 to 33 stations in the body of the lagoon from 2001 to 2004. Methodology for most of the water quality parameters was adopted from Parsons et al. (1984).

3.2 Satellite Data and Analysis

Indian Remote Sensing satellite linear imaging self scan sensor (LISS-1) data (Path/Row-20/54) on 7 April, 1989 (IRS-1A) and IRS- 1D (LISS-111) data on 16 January 2002, 18 October 2002 and 6 May, 2003 and IRS P6 data on 19 February 2005 were procured from the National Remote Sensing Agency (NRSA) in Hyderabad covering the Chilika region. The geometric registration of digital satellite data was accomplished through ERDAS Imagine Software. The digital image data were used in the ARC/INFO GIS package for further interpretation on screen. Thematic information on land use/land cover and drainage water bodies were derived from the image and final maps were generated using ARC/VIEW software packages. The spatial elements helps in detecting, delineating, identifying and evaluating the object's tone, texture, shape, size, pattern, location and association etc. Base information on infrastructure like road, railways, etc. were taken from the Survey of India Toposheets (1:250,000) and corrected with the image data. Final vector layers were prepared to generate the landuse and landcover map.

4. RESULTS AND DISCUSSION

4.1 Field Observations

The data on hydrographic and water quality parameters during summer and winter of post mouth opening conditions (2001 to 2004) have been used to assess their spatial, seasonal and interannual variability (Fig. 2-7).

4.1.1 Seasonal and interannual variability in hydrographic parameters

Surface Water Temperature: The range of temperature during summer is from 24.6°C to 35.2°C and during winter the range is 25.2°C-29.1°C. Besides the large temperature range during summer large variability in water temperature is observed from summer to winter (~10°C). Considering spatial variation in any particular season, it is observed that southern sector records the highest temperature (31.1°C) as compared to other three sectors in summer season while northern sector records the highest temperature (25.9°C) during winter (Fig. 8). Similarly, minimum temperature recorded during summer is in the outer channel (29.6°C) and in winter it is in the central sector (24.5°C).

Water temperatures do exhibit interannual variability. Summer temperature during 2004 is highest (31.5°C) as compared to average summer temperature during other three years. Highest winter temperature (26.4°C) is observed in the year 2003 (Fig. 9).

Depth: Depth of water in the lagoon shows large spatial (Fig. 2) and temporal variability (Figs. 8 and 9). Both during summer and winter, outer channel records maximum depth followed by southern, central and northern sector. Depth during winter is more than summer in all the sectors (Fig. 8) because rivers entering the lagoon drain the water after the monsoon rainfall (JJAS) with a lag of one or two months.

During summer the depth ranges from 0.35 m in the central sector to 3.2 m in the outer channel. During winter the depth ranges from a minimum (0.20 m) in the central sector to a maximum in the southern sector (3.8 m).

Considering the interannual variability in depth (Fig. 9), it is observed that 2003 winter records the highest depth (3.8 m) while the lowest depth during winter is observed in the year 2002 (1.4 m). Interannual variability during summer is less as compared to winter season. The 2001 summer records the highest depth after the opening of the new mouth in September, 2000 and subsequently summer depth reduces in the year 2002 and 2003 and shows slight enhancement in the year 2004.

Secchi Depth: Secchi depth is a measure of light attenuation in the water body and indicates the turbidity levels. Figure 3 (upper-most two panels) shows the spatial distribution of secchi depth during summer (upper panel) and winter (lower panel). It is observed that secchi depth ranges between 0.1-1.5 m during summer and 0.1-2 m during winter. Spatial variability in secchi depth is also depicted in Fig. 8 which indicates that during summer the transparency of the water is highest in the southern sector and lowest in the northern sector. The same trend in the spatial distribution is also observed during winter. For southern and central sectors the secchi depth is more during winter than during summer while the opposite trend is observed in case of northern and outer channel. It is mainly because the river runoff which enters the lagoon in the northern sector during winter carries lots of silt and sediment which finally finds way to the sea through the outer channel and thus the turbidity level in these two sectors are more during winter than during summer.

During summer season, maximum secchi depth (0.72 m) is observed during 2003 while minimum (0.54 m) is observed during 2004. During winter, highest secchi depth (0.75 m) is observed in the year 2004 and lowest (0.61 m) in the year 2002 (Fig. 9) and 2003.

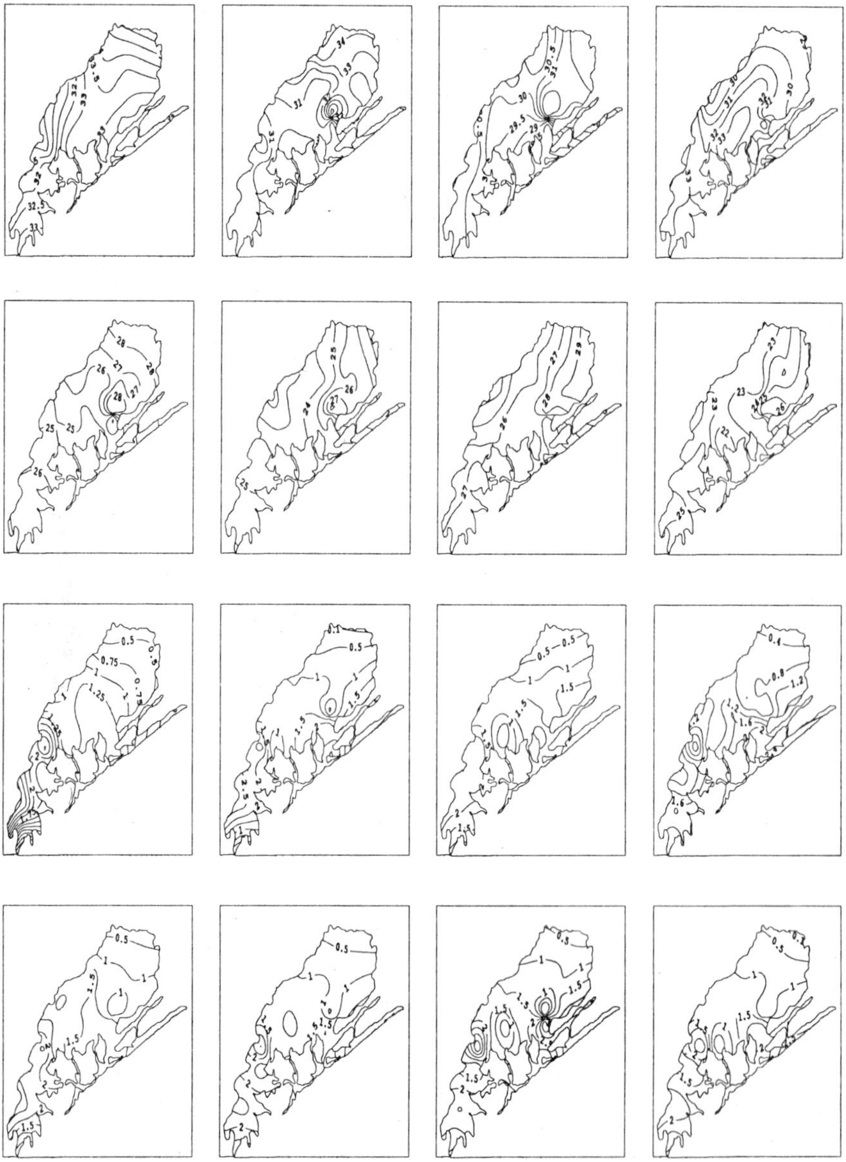

Figure 2: Water temperature (°C) during summer and winter (Panel 1 and 2 respectively); water depth (m) during summer and winter (3rd and 4th panel respectively) in Chilika lagoon during 2001-2004. (Columns 1 to 4 represent the years from 2001 to 2004.)

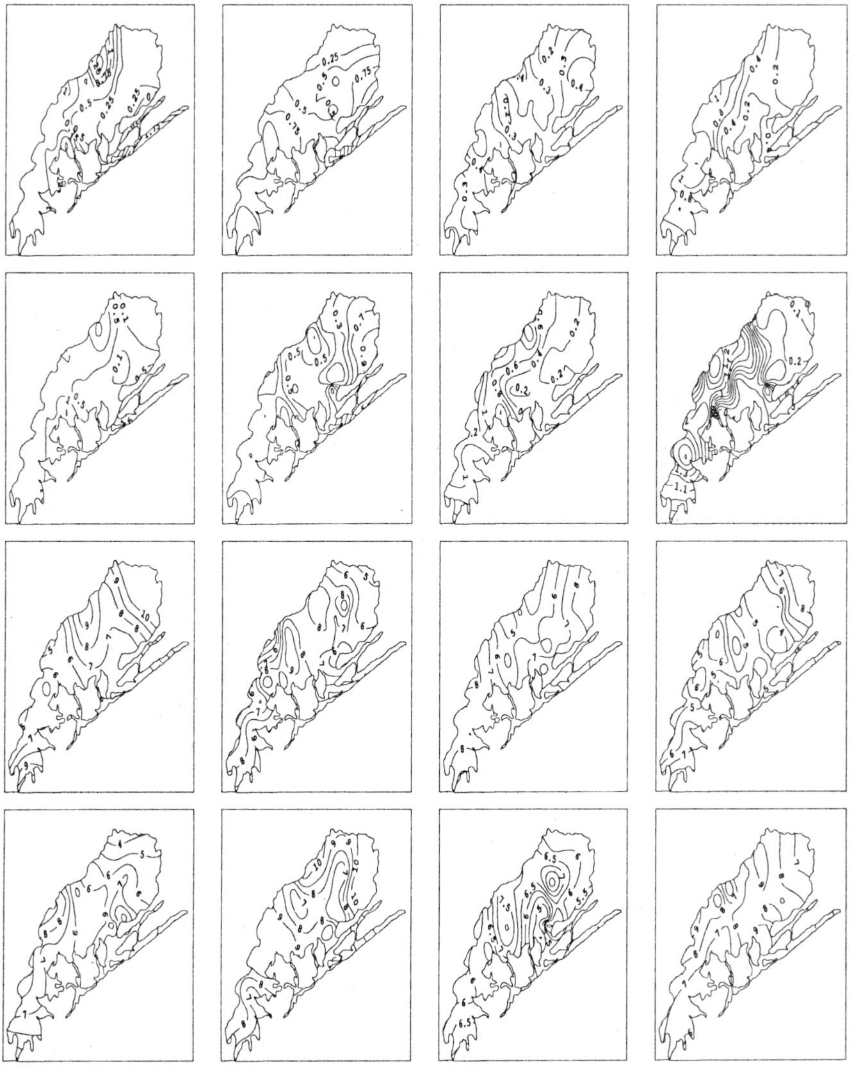

Figure 3: Secchi depth (SDD) (m) during summer and winter (Panel 1 and 2 respectively); DO (mg/l) during summer and winter (3rd and 4th panel respectively) in Chilika lagoon during 2001-2004. (Columns 1 to 4 represent the years from 2001 to 2004.)

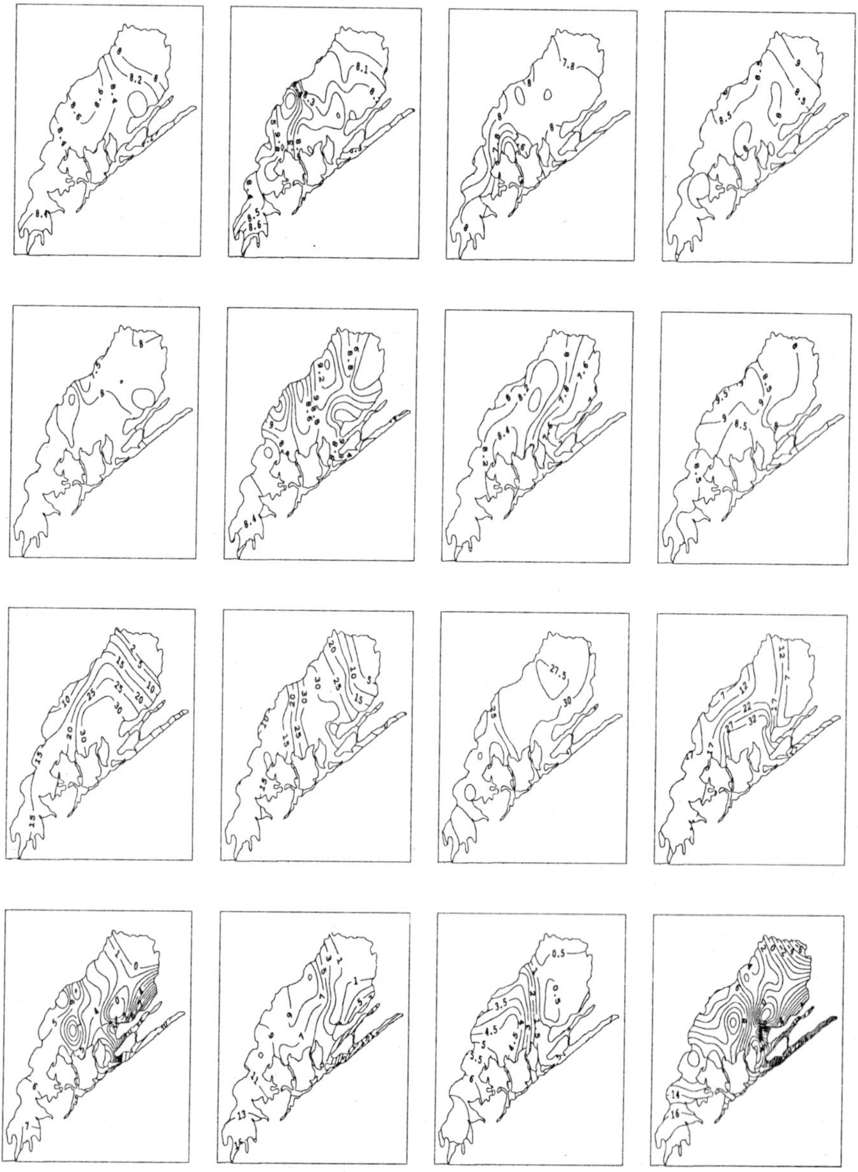

Figure 4: pH during summer and winter (Panel 1 and 2 respectively); salinity (°/oo) during summer and winter (3rd and 4th panel respectively) in Chilika lagoon during 2001-2004. (Columns 1 to 4 represent the years from 2001 to 2004.)

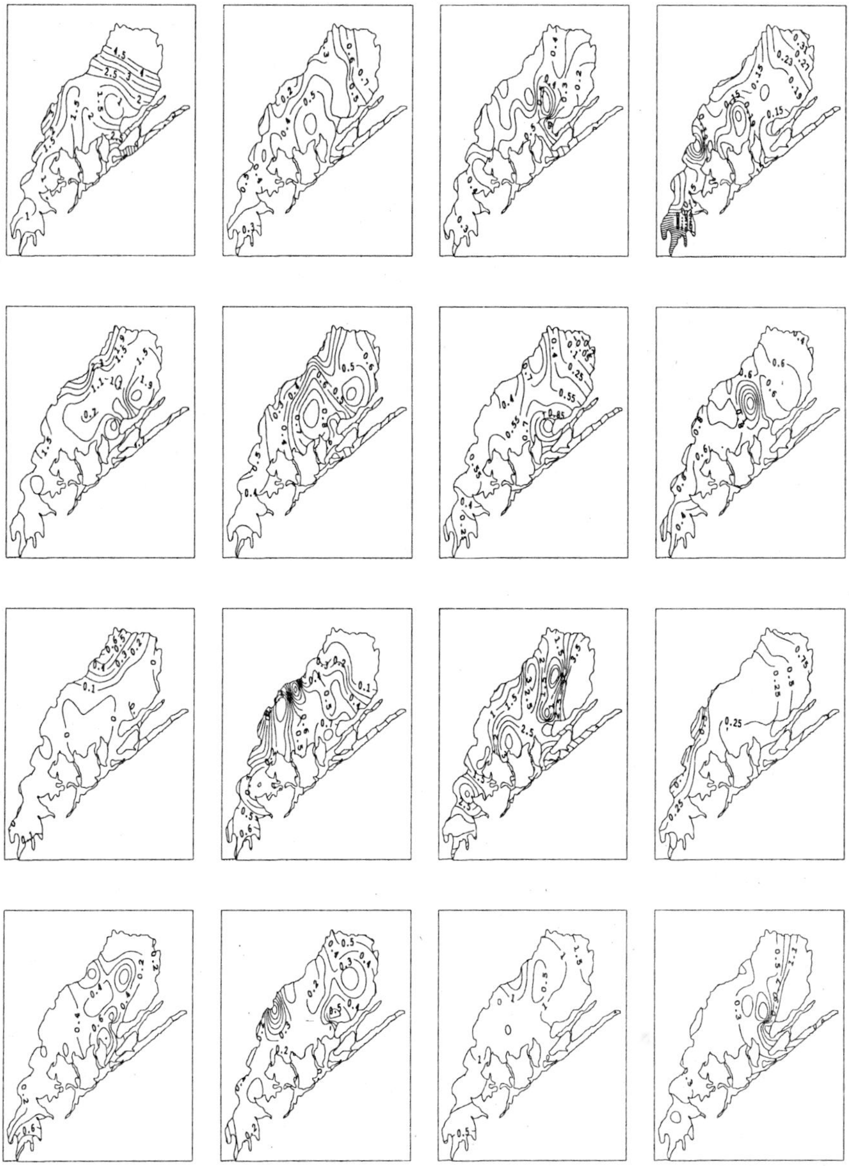

Figure 5: NH_4-N (μmol/l) during summer and winter (Panel 1 and 2 respectively); NO_2-N (μmol/l) during summer and winter (3rd and 4th panel respectively) in Chilika lagoon during 2001-2004. (Columns 1 to 4 represent the years from 2001 to 2004.)

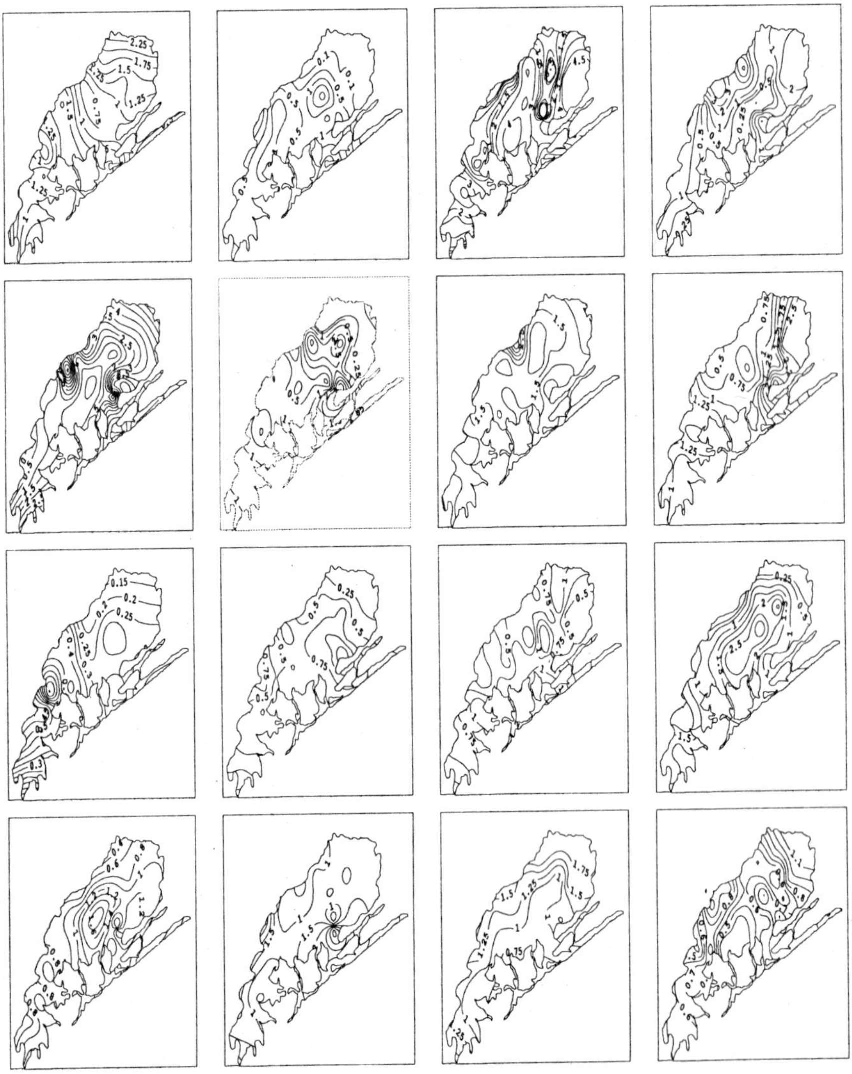

Figure 6: NO_3-N (μmol/l) during summer and winter (Panel 1 and 2 respectively); PO_4-P (μmol/l) during summer and winter (3rd and 4th panel respectively) in Chilika lagoon during 2001-2004. (Columns 1 to 4 represent the years from 2001 to 2004.)

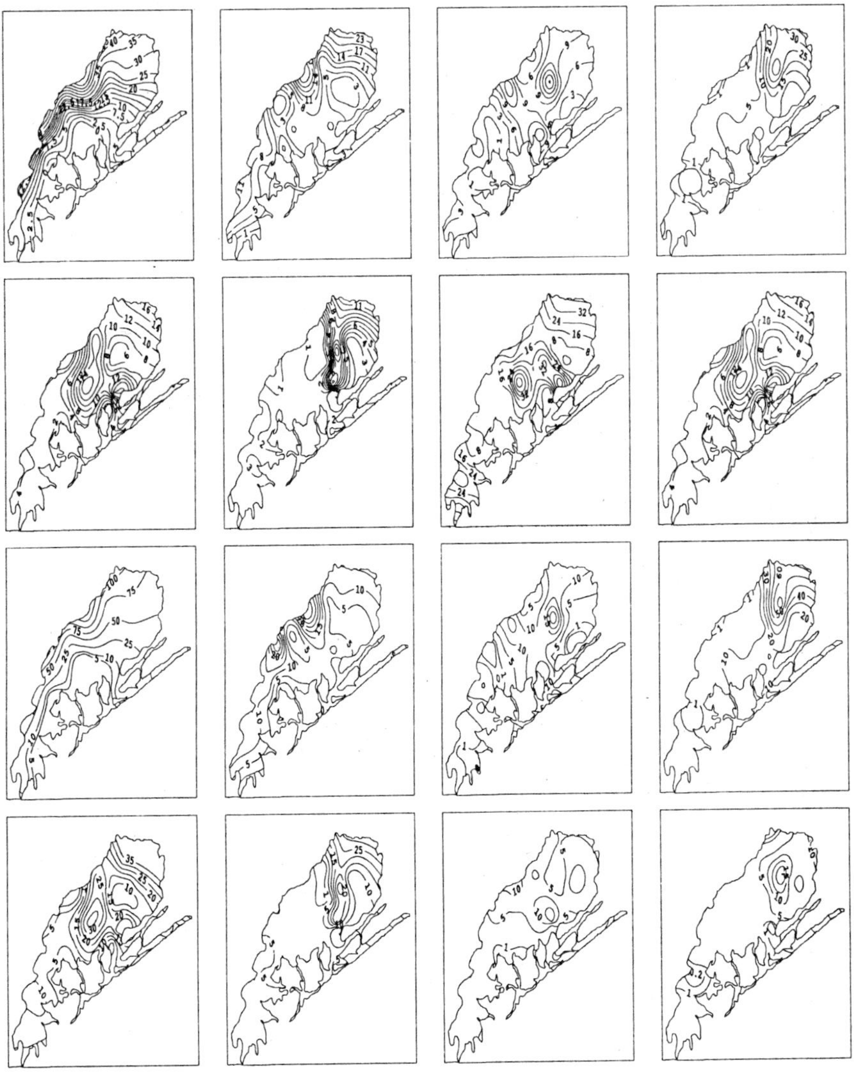

Figure 7: Chlorophyll-a (mg/m^3) during summer and winter (Panel 1 and 2 respectively); Phaeopigment (mg/m^3) during summer and winter (3rd and 4th panel respectively) in Chilika lagoon during 2001-2004. (Columns 1 to 4 represent the years from 2001 to 2004.)

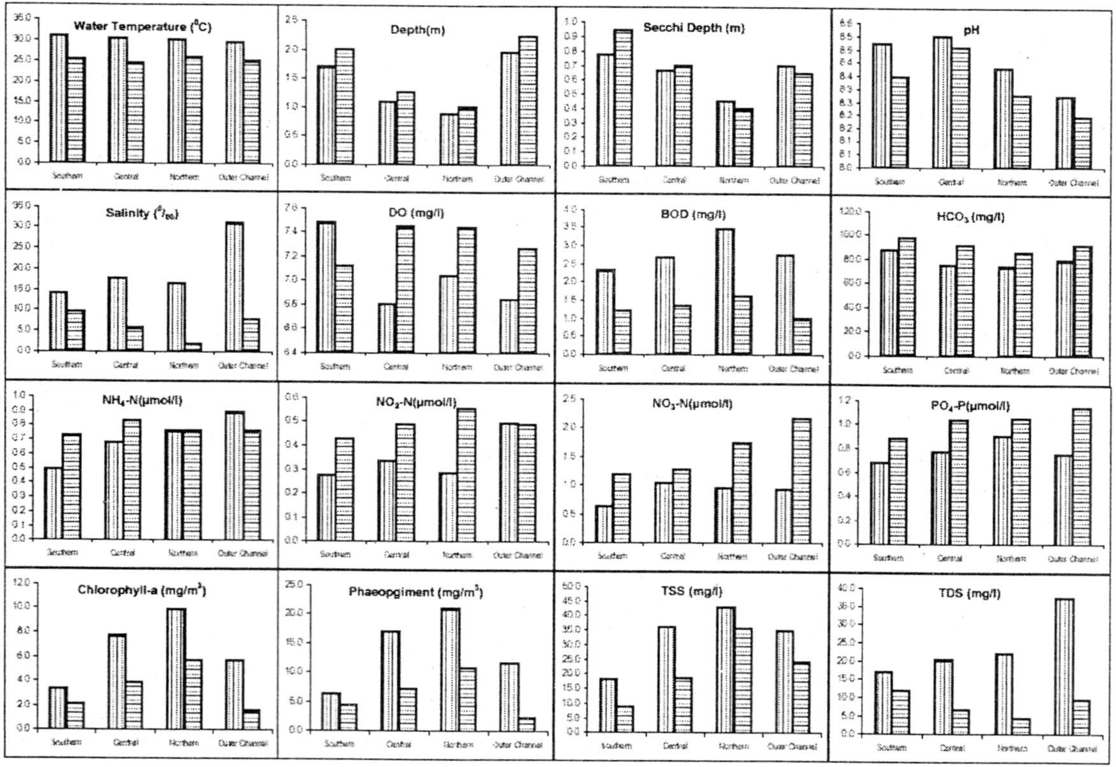

Figure 8: Seasonal variability in hydrographic and water quality parameters in different sectors of the Chilika lagoon during 2001-2004. Vertical bar represents the summer season and horizontal bar represents the winter season.

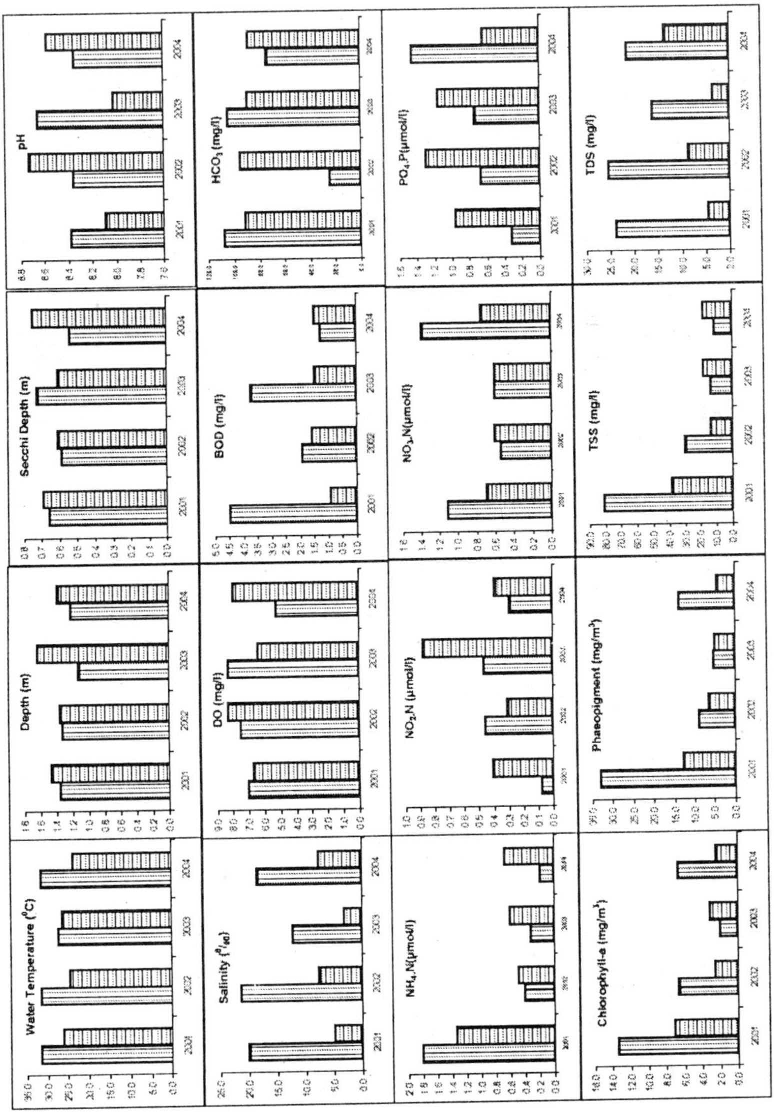

Figure 9: Seasonal and interannual variability in hydrographic and water quality parameters in Chilika lagoon during 2001-2004. Vertical bar represents the summer season and horizontal bar represents the winter season.

Salinity: Summer and winter salinity distribution in the lagoon are respectively shown in 3rd and 4th panels of Fig. 4. The salinity range during summer is observed as 2.23-34.39‰ and in winter it is 0.41-25.75‰. Summer salinity in all the sectors is distinctly more than the winter salinity (Fig. 8). During summer, outer channel records the highest salinity (30.99‰) followed by central, northern and southern sectors. But during winter, the highest salinity (9.71‰) is observed in the southern sector and lowest in the northern sector (1.98‰). Thus, the low values of salinity during winter as compared to that during summer could be attributed to the huge fresh water influx during monsoon and post-monsoon season.

Interannual variation in the salinity is also observed (Fig. 9). During summer, highest salinity is observed in the year 2002 (21.27‰) and lowest in the year 2003 (12.23‰). During winter, the highest and lowest values are respectively observed in the year 2004 (7.6‰) and 2003 (3.04‰). Therefore, it can be stated that even after opening of the new mouth, the impact is not much as the salinity variation is controlled by the fresh water influx.

4.1.2 Seasonal and interannual variability in water quality parameters

pH: Spatial distribution of pH during summer and winter are presented in the upper-most two panels of Fig. 4. The pH in the lagoon water during summer ranges between 7.95 and 9.64. Sectorwise distribution of pH during summer and winter is presented in Fig. 8. It is observed that in all the sectors summer pH values are more than the corresponding winter values. Low pH values are usually associated with low redox potential indicating higher toxicity. Thus, during winter, the toxicity level of lagoon water is more than during summer and the source could be the large influx of land and river runoff during winter. The above argument is further supported by the spatial distributional trend in the different sectors of the lagoon (Fig. 8). Outer channel and northern sector, which receive maximum freshwater, record low pH values as compared to the southern and central sectors.

The summer interannual variability is less as compared to winter (Fig. 9). Thus, winter pH which is mostly controlled by the freshwater influx can be linked to the interannual monsoon rainfall variability in the Chilika catchment. The 2003 summer records the highest pH values (8.66) while 2004 summer records the lowest pH (8.35). During winter, highest pH (8.73) is observed in 2002 and lowest in the year 2003 (80.2).

Dissolved Oxygen: Spatial distribution of dissolved oxygen in the lagoon during summer and winter are respectively shown in the 3rd and 4th panels of Fig. 3. The dissolved oxygen range during summer is 3.45-9.92 mg/l and during winter it is 4.70-9.98 mg/l. Thus, during winter, dissolved oxygen content of the lagoon water shows higher values as compared to during summer except for southern sector. However, considering the spatial distribution during summer, southern sector shows the highest value (7.47

mg/l) and the lowest (6.8 mg/l) is observed in the central sector. But during winter, spatial distributional trend is opposite with highest value in the central sector (7.44 mg/l) and lowest in the southern sector (7.12 mg/l). Thus, higher dissolved oxygen values during winter correlate with the areas of high fresh water influx and thick vegetation and agrees with the observation made by Pal and Mohanty (2002).

Summer dissolved oxygen shows highest value (8.29 mg/l) during 2003 and lowest (5.24 mg/l) during 2004 while winter dissolved oxygen shows highest (8.28 mg/l) during 2002 and lowest (6.45 mg/l) during 2003 (Fig. 9)

Bio-chemical Oxygen Demand (BOD): BOD indicates the organic pollution in water. In Chilika, the impact of organic pollution is more during summer than during winter and is apparent from Fig. 8. Thus, the organic pollution in Chilika could be associated more with weed infestation than the river and land runoff. Further, it is also observed that the magnitude of BOD during summer is highest in the northern sector (3.47 mg/l) which is considered as the most weed infested zone of the lagoon. Lowest during summer is observed in the southern sector (2.32 mg/l). During winter, northern sector shows the highest (1.61) and outer channel shows the lowest (1.00 mg/l)

Interannual variation in BOD is significant during summer as compared to the winter (Fig. 9). The 2001 summer records the highest BOD values (4.48 mg/l) while the lowest is recorded in 2004 summer (1.28 mg/l). Winter BOD values show highest in the year 2002 (1.6 mg/l) and lowest in the year 2001 (0.91).

Bicarbonates (HCO_3): Carbonates play important role in neutralization of strong bases and acids. The chemistry of many important reactions is affected by the presence of carbonate species. Winter HCO_3 values are more than the summer HCO_3 values in all the sectors of the lagoon (Fig. 8). Both during summer and winter, highest and lowest HCO_3 values are observed in southern and northern sectors respectively. The interannual variability in bicarbonates is maximum during summer and minimum during winter (Fig. 9). The 2001 summer records the highest values and 2002 summer records the lowest values. During winter the highest is observed in the year 2002.

Water Nutrients: Nitrite (NO_2-N), nitrate (NO_3-N), ammonium (NH_4-N) and phosphate (PO_4-P) are important water nutrients which have been discussed here. Figures 5 and 6 present the spatial distributional trend of the above nutrients. In Chilika lagoon water during summer, the ranges of nitrite, nitrate, ammonium and phosphate are respectively observed as 0.-2.08 μmol/l, 0.01-5.17 μmol/l, 0.08-4.26 μmol/l and 0.17-3.37 μmol/l. During summer, the concentration of nutrients in the lagoon water is less than the winter concentration in all the four sectors of the lagoon (Fig. 8). Further, factor analysis shows the negative correlation between salinity and nutrient during winter which strongly indicates the source of nutrients as the river and land runoff. However, during summer, no relationship could be established between salinity and nutrients. During summer, more nitrite concentration is in the outer channel as compared to other sectors which suggests the contribution due to sea water. Lowest nitrite concentration during summer is observed in the southern sector and this further corroborates the role of sea water influx in the nitrite concentration during summer. Nitrate concentration during summer shows very less spatial variability whereas during winter the highest concentration is observed in the channel followed by northern, central and southern sector. Phosphate concentration shows similar trends as nitrate (NO_3-N) and thus could be due to same sources during summer and winter. Ammonium concentration during summer shows a constant increase from southern sector to central, northern and outer channel whereas during winter the spatial variability is very less.

Interannual variability in water nutrients during summer and winter is shown in Fig. 9. Except for some years with higher concentration of nutrients during summer, winter nutrients concentration are more in most of the years.

Chlorophyll-a and Phaeopigments: The spatial distribution during summer and winter are shown in Fig. 7. The range of chlorophyll-a in summer is 0.31-54.04 mg m^{-3} and in winter it is 0.09-18.14 mg m^{-3}. Figure 8 presents the distribution of chlorophyll-a during summer and winter in different sectors. It is observed that summer values are distinctly higher than winter values in all the sectors. However, highest values during summer are observed in the northern sector (32.47 mg m^{-3}) and lowest in southern sector (3.34 mg m^{-3}). Despite the fact that nutrient concentrations are high during winter, higher chlorophyll concentrating during summer, having no significant relationship with salinity, could be attributed to the luxuriant growth of vegetation/weed cover during summer. Winter chlorophyll-a concentration also shows highest in the northern sector (5.70 mg m^{-3}) and lowest in the channel (1.56 mg m^{-3}).

Interannual variability in chlorophyll-a concentration is significant (Fig. 9) and shows that both during summer and winter; concentration was highest in the year 2001, following the opening of the new mouth in September 2000. However, a gradual decline is observed thereafter but with a slight enhancement in 2004 summer. Phaeopigment concentration shows almost similar trend as the chlorophyll in spatial, seasonal and interannual variability.

Total Suspended Solid (TSS): Figure 8 shows the spatial distribution of TSS during summer and winter. It is observed that TSS values are more during summer than those during winter. Despite the fact that TSS load is more during winter following the huge river and land drainage, the high summer concentration could be attributed to the churning of the water due to wind action which is more prominent in shallow water bodies like Chilika. Higher TSS concentrations, both during summer and winter, in the northern sector, the shallowest among all the sectors, clearly indicate the role of wind churning in TSS concentration. Southern sector records lowest TSS concentration during summer and winter due to its higher depth and less wind churning.

Interannual variability in TSS concentration is quite significant and shows a constant declining trend from 2001 to 2004 during summer. However, during winter, the declining trend changes in the year 2003 and 2004 (Fig. 9).

Total Dissolved Solids (TDS): TDS concentration in the lagoon shows significant spatial variability during summer and winter (Fig. 8). The highest concentration during summer is observed in the outer channel followed by northern, central and southern sector. This clearly indicates the role of sea water influx in controlling the TDS concentration. Further, the factor analysis also shows significant positive correlation between salinity and TDS and hence supports the view that sea water influx is the major contributor of TDS. During winter, highest concentration is observed in southern sector with highest salinity and lowest concentration in the northern sector with lowest salinity. Interannual variability in TDS is more during winter than during summer (Fig. 9). The 2002 summer shows the highest TDS value while 2003 shows the lowest TDS value. On the other hand 2004 winter shows highest TDS concentration and 2003 winter shows the lowest TDS concentration.

4.2 Statistical Analysis of Hydrographic and Water Quality Parameters

Multivariate statistical methods including factor analysis have been successfully applied in hydrochemistry for many years. Recent studies give emphasis on the application of principal component analysis/factor analysis for water quality assessment (Spencer, 2002; Praus, 2005; Boyacioglu, 2006). The usefulness of the statistical analysis lies with deriving hidden information from the data set about the possible influences of the environment in water quality (Spanos et al., 2003).

Factor analysis attempts to explain the correlations between the observations in terms of the underlying factors which are not directly observable (Yu et al., 2003). There are three stages in factor analysis (Gupta et al., 2005) which have been adopted in the present study. They are:

i. For all the variables a correlation matrix is generated.
ii. Factors are extracted from the correlation matrix based on the correlation coefficients of the variables by the centroid method.
iii. To maximize the relationship between some of the factors and variables, the factors are rotated by varimax roation (Ahmed et al., 2005).

4.2.1 Factor Analysis

Based on the above method three significant factors were generated which explain 52.2% of the variance in data set and factor loading are given in Table 2 for summer and winter season.

Table 2: Factor loadings of hydrographic and water quality parameters in Chilika Lagoon during summer and winter

Summer	*Winter*
Number of variables: 15	Number of variables: 15
Method: Principal components	Method: Principal components
log(10) determinant of correlation matrix: *2.5049*	log(10) determinant of correlation matrix: *2.6807*
Number of factors extracted: 3	Number of factors extracted: 3
Eigenvalues: *3.09121 3.02691 1.70654*	Eigenvalues: *4.48274 2.10719 1.39741*
Marked loading are > 0.50	Marked loading are > 0.50
Factor rotation: Varimax normalised	Factor rotation: Varimax normalised

	Factor 1	*Factor 2*	*Factor 3*	*Factor 1*	*Factor 2*	*Factor 3*
WT (°C)	0.389	0.038	-0.355	-0.741	-0.275	0.195
Depth (m)	0.305	-0.064	0.485	0.247	-0.599	0.212
SDD (m)	-0.365	-0.110	0.522	0.597	-0.462	0.055
TSS (mg/l)	0.192	0.725	-0.122	-0.713	0.131	-0.438
TDS (mg/l)	0.864	0.059	0.144	0.743	-0.243	-0.297
pH	-0.775	0.023	0.194	0.673	0.457	-0.019
Salinity (‰)	0.900	0.003	0.114	0.728	-0.436	-0.177
DO (mg/l)	-0.226	0.309	0.683	0.466	0.648	-0.134
BOD (mg/l)	-0.110	0.792	0.285	0.089	0.560	0.032
HCO_3-(mg/l)	-0.342	0.486	-0.011	0.267	-0.233	0.316
NH_3-N(mg/l)	0.178	0.836	-0.056	-0.404	-0.372	-0.476
NO_2-N(mg/l)	0.119	-0.398	0.282	-0.502	-0.094	0.439
NO_3-N(mg/l)	-0.122	0.113	-0.400	-0.576	-0.108	-0.139
PO_4-P(mg/l)	0.161	-0.549	-0.456	-0.160	0.190	0.632
Chl-a (mg/m^3)	0.285	0.608	-0.373	-0.636	0.024	-0.210
Eigenvalues	2.920	3.048	1.857	4.483	2.107	1.397
% of total variance	20.608	20.179	11.377	29.885	14.048	9.316
Cumulative % of variance	20.608	40.787	52.164	29.885	43.933	53.249

Summer

Factor 1: This factor accounts for 20.6% of the total variance. High positive loading of total dissolved solids and salinity and negative loadings of pH suggest that seawater influx mainly controls the salinity and dissolved solids. Negative factor loading of pH explains the disproportion between this parameter and factor one which mostly represents the process of influx of seawater. High negative loading of pH indicates that with the influx of sea water, nitrification and decomposition of organic matter occur which results in lowering of pH.

Factor 2: Factor 2 accounts for 20.2% of the total variance. High positive loading of Total Suspended Solid (TSS), Biological Oxygen Demand (BOD), NH_4-N and moderated positive loading of chlorophyll-a are associated with this factor. High positive loading of the first three factors suggests that river run-off and land run-off/terrestrial input are the main sources of suspended solid and nutrient concentration in coastal lagoon. Further, it is indicated that land run-off from the western catchment and river run-off which are represented through Factor 2 are the main sources of organic pollution (Boyacioglu, 2006) in the lagoon.

Factor 3: This factor accounts for only 11.4% of the total variance and is represented through positive loading of dissolved oxygen associated with moderately positive loading of secchi disk depth. Thus, the factor represents the high productivity status due to photosynthetic activities of phytoplankton and higher solubility of oxygen in low saline/brackish water Chilika lagoon.

Winter

Factor 1: For winter season the factor loadings are shown in Table 2. It is observed that total dissolved solids and salinity have high positive loadings associated with high negative loadings of TSS and moderately negative loadings of chlorophyll-a and nutrients (NO_3-N, NO_2-N and NH_4-N). This factor explains the role of seawater influx which is positively correlated with salinity and dissolved solid and negatively correlated with total suspended solids and nutrients which are mostly of land origin or from the river run-off.

Factor 2: This factor explains 14.1% of the total variance and is represented by the positive loading of DO which represents the process of high productivity. The negative loadings of salinity associated with positive loadings of BOD and DO further suggest that the input of organic waste is through river runoff (Panigrahy et al., 1999) and the productivity is favoured in a low saline environment.

Factor 3: This factor accounts for only 9.3% of the total variance and is represented by the only positive loading PO_4-P. Thus, the factor indicates that the river runoff or land drainage is the chief source of nutrients in the coastal lagoon.

4.3 Spectral Analysis

It is well known that remotely sensed data acquired from the Indian Remote Sensing satellites have the potential to provide meaningful information on the water quality, silt load, productivity and vegetation cover of lagoons/ lake, estuaries and other coastal water bodies with a cost effective procedure and better synopticity unobtainable using conventional methods. Witzig and Whitehurst (1981) provided an excellent review of the literature describing the application of remote sensing to surface water quality studies. Therefore in the present study attempts have been made to monitor the water quality and vegetation coverage and to estimate their time rate of change by determining the areas covered under different water and vegetation types.

4.3.1 False Colour Composite (FCC)

False colour composites are obtained by combining the images of band-4, band-3 and band-2, which are respectively assigned red, green and blue colours. Emergent vegetation are shown in red, water in deep blue to light blue and the black colour depicts under water/submerged vegetation inside the lagoon. False colour composite for summer and winter are discussed with representative data sets.

The FCC during summer (Fig. 10a) depicts the areas of emergent vegetation (red), submerged vegetation (black), deep water (deep blue) and shallow water (light blue) regions and fairly agree with field observations. Nalabana Island during summer is very distinct with reddish tinge and suggests that the Island is dry during summer and covered with sparse emergent vegetation. Nalabana Island is completely submerged and the submerged vegetation (dark colour) is distinctly shown. During monsoon and post monsoon seasons large influx of fresh water to the lagoon occurs through the rivers (Fig. 1). Therefore, the water quality and vegetation cover changes as compared to summer. Figure 10b depicts the features of Chilika lagoon during winter. The areas of emergent and submerged vegetation, deep and shallow water regions are very well delineated in the FCC. It is very much distinct that the vegetated area (emergent and floating) during winter is less as compared to summer. However, submerged vegetation during winter is more than that during summer. The area statistics are shown in Table 3. The area statistics show that deep (blue) and shallow water (light blue) bodies are relatively less during summer than those during winter and fall. The Nalabana Island is partly submerged during winter with emergent vegetation (red) at the periphery and submerged vegetation (black) at the core of the Island.

4.3.2 Classified Images

Satellite data sets representing summer and winter were digitally analyzed and the maximum likelihood classification was used to obtain spectral

Table 3 : Tree structure of the areas (km^2) under major spectral classes and total water area of Chilika lagoon for the four satellite passes during post-mouth opening conditions.

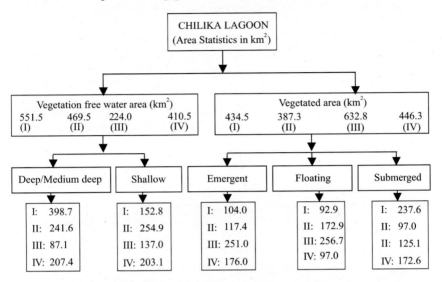

CHILIKA LAGOON
(Area Statistics in km^2)

Vegetation free water area (km^2)			
551.5 (I)	469.5 (II)	224.0 (III)	410.5 (IV)

Vegetated area (km^2)			
434.5 (I)	387.3 (II)	632.8 (III)	446.3 (IV)

Deep/Medium deep	Shallow	Emergent	Floating	Submerged
I: 398.7	I: 152.8	I: 104.0	I: 92.9	I: 237.6
II: 241.6	II: 254.9	II: 117.4	II: 172.9	II: 97.0
III: 87.1	III: 137.0	III: 251.0	III: 256.7	II: 125.1
IV: 207.4	IV: 203.1	IV: 176.0	IV: 97.0	IV: 172.6

(I: January, 2002, II: October, 2002, III: May, 2003 and IV: February, 2005)

signature of different classes during summer (May 2003) and winter (January 2002) seasons in the post mouth opening period.

Six different classes during summer (Fig. 10c) and winter (Fig. 10d) were obtained and the areas covered under different classes were estimated. The six classes are deep water, shallow water, emergent vegetation, free floating vegetation, submerged vegetation and inland vegetation. Excluding the inland vegetation, the changes that are apparent from winter (January 2002) to next year summer (May 2003) are decrease in the area of deep water, shallow water and submerged vegetation while there is significant increase in the emergent and free floating vegetation. Thus, the remarkable changes from summer to winter are reduction in the vegetation free water area and enhancement in the vegetated area. The results partly agree with the observation of Pal and Mohanty (2002) for the pre-mouth opening period which showed minimum vegetation free water area and maximum vegetated area during summer.

4.4 Landuse/Landcover

Spatial distribution of different landuse/landcover pattern of the Chilika and its catchments area covering 2724.701 sq. km. has been assessed from IRS 1A LISS-I data on 7 April, 1989 and IRS 1D LISS-III data on 16 January, 2002. The important landuse/landcover patterns have been identified and are presented in Figs. 10e and 10f following Mishra (2004). Table 4 presents the area statistics of the different landuse/landcover patterns and also delineates the changes in the area from 1989 to 2002.

Built-up land: Built-up land includes the settlements in the villages including vegetation around it. It also includes the urban areas of Balugaon and Rambha

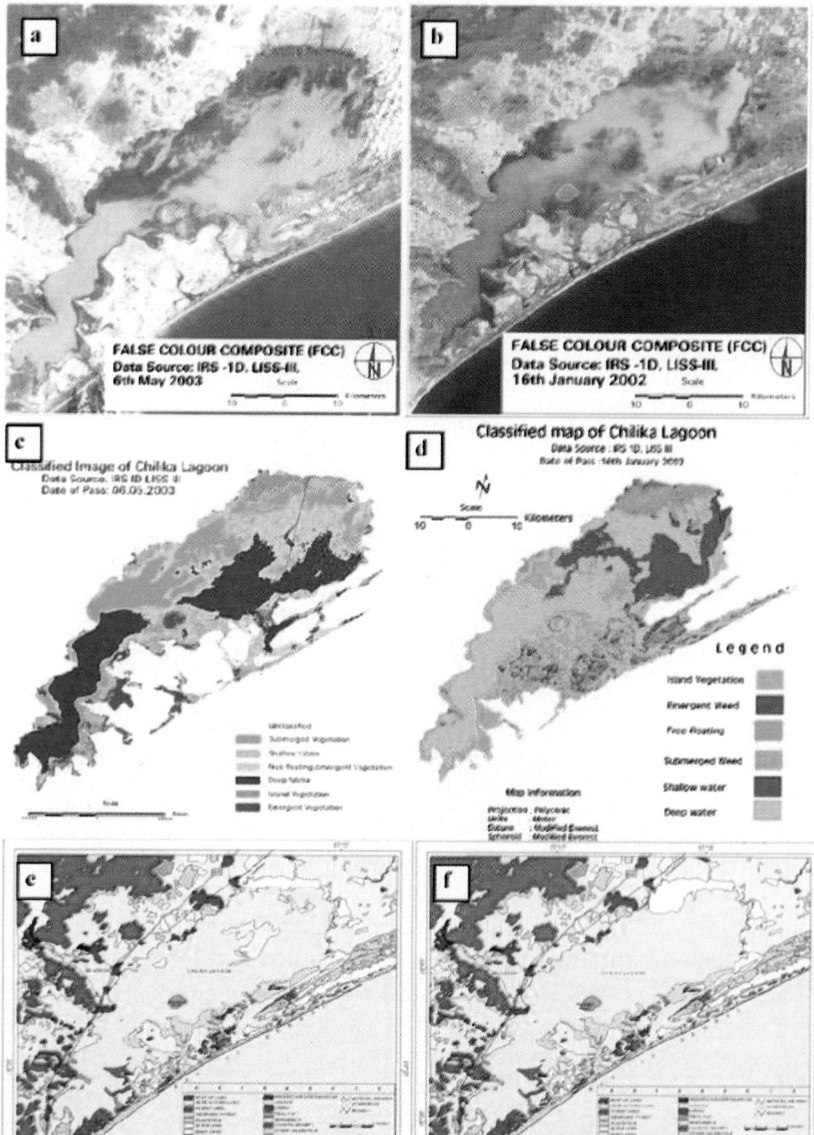

Figure 10. (a) FCC Image (May 2003); (b) FCC Image (January 2002); (c) Classified Image (May 2003); (d) Classified Image (January 2002); (e) Landuse/landcover map (April 1989) and (f) Landuse/landcover map (January 2002).

Table 4: Statistics of Landuse/landcover pattern of Chilika surroundings

Category	*Area IRS 1A LISS 1 07.04.1989 (sq.km)*	*Area IRS 1D LISS III 21.05.2002 (sq.km)*	*2002-1989*
Built-up land	53.825	56.831	3.006
Agriculture	778.258	784.221	5.963
Forest Area	348.128	350.081	1.953
Degraded Forest	32.783	30.776	-2.007
Plantation	127.347	112.91	-14.437
Scrub land	232.676	205.651	-27.025
Coastal Sandy area	0.29	0.24	-0.05
Rocky Area	2.357	2.37	0.013
Water logged area	8.112	8.098	-0.014
Swampy/marshy area (Inland)	59.917	75.746	15.829
Reservoir/Tank/Pond	9.058	9.06	0.002
Lagoon (Water Spread)	909.38	959.807	50.427
Creek	0.413	0.303	-0.11
Tidal/spit	9.217	11.647	2.43
Beach/Spit	7.97	7.242	-0.728
Coastal Swampy Area	74.712	41.76	-32.952
Agriculture Pond	20.961	58.62	37.659
Islands	8.755	8.835	0.08
Weed Infested area	40.542	0.503	-40.039
	2724.701	2724.701	0

Notified Area Councils. In the satellite image the village and urban settlements are seen as clusters of discrete red patches due to the presence of green cover in the form of orchards. However, the settlement areas were further refined by incorporating the information from Survey of India topographical maps onto the classified data.

Agriculture Land: Areas covered under kharif and rabi crops are included in this category. Major kharif crop is paddy. Rabi crops are seen along the rivers and channels. Vast areas of crop land are seen on the palaeo-mudflats in the north and north-eastern sector of the lagoon. Croplands are also delineated in the barrier islands where people practice cultivation after putting embankments to check the salinity. Such areas are seen in Parikud, Malud and Krushna Prasad Garh area (see Fig. 1).

FOREST

Deciduous forest: Deciduous forests are mostly confined in the hills bordering north-west and western side of the Chilika lagoon. The common species met with are *Shorea robusta, Mangifera indica, Eugenia jambolana, Schleichera trijuga, Terminalia tomentosa, Dendrocalamus strictus, Madhuca latifolia, Tamarindus indica, Bombax malabaricum* etc.

Plantation: Forest plantations like eucalyptus, casuarinas, acacia, bamboo etc. are seen in the forest plantation areas. They are seen on the hill slopes, lateritic uplands and barren strip lands. Casuarinas plantation is seen on the sand dunes along the spit and sand ridges. Agricultural plantations include cashew, coconuts etc. Cashews with mixed species are seen on the banks along the outer channel and Satapada region.

WASTELAND

Land with/without scrub: Rolling uplands or undulating upland have scrubs on them and found in the north-west, west and southern portion of the lagoon. The species composition of the scrubs is of mixed type.

Sandy area (coastal): Barren sandy areas adjacent to the coastal plain are included in this category. They mostly comprise sand cast areas and flattered sand dunes.

Barren rocky/stony waste/sheet rock area: Barren/stony wastelands are seen on the western and southern border of the lagoon.

WETLANDS

Inland Wetlands (Natural)

Waterlogged (seasonal): The low-lying areas in the palaeo-mudflats on the north and north-eastern border are categorized as waterlogged areas. These are formed on back swamps and adjacent to the rivers-Daya, Bharagavi and Makara.

Swamp/marsh: Swamp/marsh areas are seen on the water logged areas and other low relief areas. The palaeo-mudflats on the north and north-eastern border of the lagoon show several cut-off meanders, which were once part of the drainage system flowing to the lagoon.

Coastal Wetlands (Natural)

Tidal flat/Mud flat: The tidal flat/mud flats are seen in the channel zone of Bhubania nadi near Magarmukh area and inter-tidal regions of the islands/bars. They are ideal habitation for several species of benthic fauna. Around 9.217 sq. km. and 11.647 sq. km of area have been delineated inside the lagoon in 1989 and 2002 respectively. The palaeo-mudflats on the north and north-eastern border of the lagoon show several cut-off meanders, which were once part of the drainage system flowing to the lagoon.

Sand/Beach/Spit/Bar: Barren sandy areas along the coast, spit and sandy bar are estimated to cover 7.97 sq. km and 7.242 sq. km of area in 1989 and 2002 respectively. The bars/sandy areas are occupied by beach vegetation.

Marshy/Swampy area: Vegetation growth is very acute on the north, north-eastern and north-western sector of the lagoon. Due to increase in siltation from the river drainage into the lagoon, the mouth areas become congenial for growth of such marsh vegetation. They cover around 59.917 sq. km and 75.746 sq. km of area in 1989 and 2002 respectively and their growth becoming alarming.

Other Islands: The Nalaban Island inside the lagoon is a bird sanctuary. Major vegetation species of the area include *Phragmites karka*, *Potamogeton pectinatus*, *Halophila ovalis*, *Hydrilla verticillata*, *Ceratophyllum demersum*, *Azolla pinnata* etc. Besides Nalabana, other small islands having area of 8.835 sq.km in 1989 and 8.835 sq.km in 2002 are also present inside the lagoon.

Coastal Wetlands (Man-made)

Aquaculture ponds: Barren mudflats, inter-tidal, sheltered zones in the outer channel area are converted to aquaculture ponds in Satapada, Brahmagiri, Krushnaprasad Garh areas. The areas covered under such wetlands are estimated as 20.961 sq. km and 58.62 sq. km respectively in 1989 and 2002. Small sporadic aquaculture ponds could not be delineated due to scale limitations.

5. CONCLUSIONS

The opening of the artificial mouth on 23 September, 2000 is one of the major steps taken for the conservation and restoration of the ecosystem of the Chilika lagoon. Following the new mouth opening the ecological changes in lagoon environment were monitored through field observations and using Indian remote sensing satellite data during summer and winter period. Hydrographic and water quality parameters in the lagoon show significant spatial and temporal variability. After opening of the new mouth, outer channel records highest depth followed by southern, central and northern sectors. Similarly other parameters show distinct spatial distributional trend. However, the seasonal variability is remarkable and dominates over inter-annual variability. In all the four sectors of the lagoon, pH, salinity, BOD, chlorophyll-a, phaeopigment and TSS are more during summer than those during winter while depth, dissolved oxygen and water nutrients show an opposite trend. Thus, the distinct seasonal variability in hydrographic and water quality parameters associated with spatial variability allow lots of gradient to exit, both in time and space, in the lagoon environment. Existence of such gradient in a lagoon helps maintain its biodiversity and increased productivity. Statistical analysis of the observed data reveals the major processes which are in operation and their significant association with specific water quality and hydrographic parameters. Remote sensing observation and the data analysis further helps in detecting the changes in the lagoon environment in the time-space continuum. As in the pre-mouth opening period, summer season in the post-mouth opening period also records more vegetated area and less vegetation-free water area. It is therefore apparent that large seasonal variability in lagoon environment in conjunction with the seasonal variability in freshwater influx into the lagoon, seasonal/annual cycle of vegetation growth and decay and nutrient cycle are some of the important components of the contemporary phases of lagoon transformation.

ACKNOWLEDGEMENTS

One of the authors (USP) acknowledges the financial assistance offered by the Council of Scientific and Industrial Research (CSIR), Government of India in terms of Senior Research Fellowship. Part of the work reported in this paper was also supported by Department of Ocean Development (DOD), Government of India through a sponsored research project.

REFERENCES

Ahmed, S., Hussain, M. and Abderrahaman, W., 2005. Using multivariate factor analysis to assess surface/logged water quality and source of contamination at a large irrigation project at Al-Fadhi, Eastern Province. Saudi Arabia. *Bull. Eng. Geol. Env.*, **64:** 315-323.

Barnes, R.S.K., 1980. Coastal Lagoons. Cambridge studies in Modern Biology, 1, 105p.

Boyacioglu, H., 2006. Surface water quality assessment using factor analysis. *Water SA.*, **32:** 389-393.

Chilika, 2001. Chilika—A new lease of life. Chilika Development Authority, Bhubaneswar, Orissa. pp 13.

Fisher, T.R., Peele, E.R., Peele, J.W., Ammerman, J.W. and Harding, L.W., 1972. Nutrient limitation of phytoplankton in Chesapeake Bay. *Mar. Ecol. Progr. Ser.*, **82:** 51-63.

Gupta, A.K., Gupta, S.K. and Patil, R.S., 2005. Statistical analyses of coastal water quality for a port and harbour region in India. *Environ. Monit. Asses.*, **102:** 179-200.

Mishra, B.N., 2004. Landuse/Landcover studies of Chilika lagoon, east coast of India—A Remote Sensing and GIS approach. Berhampur University, Master of Philosophy's thesis, 29p.

Mohanty, P.K., Pal, S.R. and Mishra, P.K., 2001. Monitoring ecological conditions of a coastal lagoon using IRS Data: A case study in Chilika, East Coast of India. *Journal of Coastal Research, Special Issue* **34:** (ICS 2000 Proceedings) 459-469.

Nichols, M. and Allen, G., 1981. Sedimentary processes in coastal lagoons. *Coastal Lagoon Research, Present and Future*. UNESCO Technical Papers in Marine Science, **33:** 27-80.

Pal, S.R. and Mohanty, P.K., 2002. Use of IRS-1B data for change detection in water quality and vegetation of Chilika lagoon, east coast of India. *International Journal. Remote Sensing*, **23(6):** 1027-1042.

Panigrahi, S.H., 2005. Seasonal variability of phytoplankton productivity and related physico-chemical parameters in the Chilika lake and its adjoining sea. Berhampur University, Doctor of Philosophy's thesis, 258p.

Panigrahy, P.K., Das, J., Das, S.N. and Sahoo. R.K., 1999. Evaluation of the influence of various physico-chemical parameters on coastal water quality, around Orissa, by factor analysis. *Ind. J. Mar. Sci.*, **28:** 360-364.

Parsons, T.R., Maita, Y. and Lalli, C.M., 1984. A manual of Chemical and Biological Methods for Seawater Analysis. Pergamon Press, New York, 173p.

Phleger, F.B., 1981. A review of some general features of coastal lagoons. *Coastal Lagoon Research, Past, Present and Future*, UNESCO Technical Papers in Marine Science, **33:** 7-14.

Praus, P., 2005. Water quality assessment using SVD-based principal component analysis of hydrological data. *Water SA.*, **31:** 417-422

Spancer, K.L., 2002. Spatial variability of metals in the intertidal sediments of the Medway Estuary, Kent, UK. *Mar. Pollut. Bull.*, **44:** 933-944.

Spanos, T., Simeonov, V., Stratis, J. and Xristina, X., 2003. Assessment of water quality for human consumption. *Microchim. Acta.*, **141:** 35-40.

Witzig, S. and Whitehurst, C., 1981. Literature review of the current use and technology of MSS digital data for lake tropic classification. *In:* Proceedings of the 1981 Fall Meeting of the American Society of Phtotogrammetry, San Francisco, pp.1-20.

Yu, S., Shang, J. and Guo, H., 2003. Factor analysis and dynamics of water quality of the Songhua River Northeast China. *Water, Air Soil Pollut.*, **144:** 156-169.

Assessment and Monitoring the Coastal Wetland Ecology Using RS and GIS with Reference to Bhitarkanika Mangroves of Orissa, India

Chiranjibi Pattanaik, Ch. Sudhakar Reddy, M.S.R. Murthy and D. Swain

Forestry & Ecology Division
National Remote Sensing Agency, Hyderabad 500037, India
julu2k@yahoo.com

1. INTRODUCTION

Mangrove forest is a vegetation community formed by a variety of salt-tolerant species growing in the inter-tidal areas and estuary mouths between land and sea. Mangrove forests are one of the most productive wetlands on the earth. It can provide critical habitat for a diverse marine and terrestrial flora and fauna. Yet, these unique coastal tropical forests are among the most threatened habitats in the world. Traditionally, local communities in mangrove ecosystems collected fuel wood, harvested fish and other natural resources (Bandarnayake, 1998; Dahdouh-guebas et al., 2000a). However, in recent decades many coastal areas have come under intense pressure from rapid urban and industrial development, compounded by a lack of governance or power among environmental institutions. Mangroves have been overexploited or converted to various other forms of land use, including agriculture, aquaculture, salt ponds, terrestrial forestry, urban and industrial development and for the construction of roads and embankments (Das et al., 1997). Mangroves can be affected by several different activities simultaneously, or over time as landuse patterns change (Dahdouh-Guebas, 2000b).

The mangrove forests comprise 15.8 million hectares, roughly less than half the original mangrove forest cover and are fast declining further at an assumed rate of 2 to 8% per year in the world (Spalding, 1997). India has a total area of 4461 sq. km. under mangroves which is 0.14% of the country's total geographic area. It accounts for about 5% of the world's mangrove vegetation (FSI, 2003). Nearly 57% mangroves are found along the east coast. The National Remote Sensing Agency (NRSA) recorded a decline of 59.18 sq. km. of mangrove between 1972-1975 and 1980-1982 (NRSA, 1983). According to the Government of India report (1987), India lost 40% of its mangrove area during the last century (Kumar, 2000).

Mangroves have not received proper attention and they have been subjected to over-exploitation and encroachment and hence there is a need for conservation and management of mangrove forests. Recognizing the environmental, social and economic importance of mangrove ecosystems, a National Mangrove Committee was set up in the Department of Environment, Forest & Wildlife in 1976, for a close monitoring of these valuable coastal resources and to evolve a management plan for protection of mangroves (Mangrove Status Report, 1987). Several floristic studies have been done in Bhitarkanika Wildlife Sanctuary by several botanists from time to time (Banerjee, 1984; Banerjee et al., 1990; Saxena and Brahmam, 1996; Chadha and Kar, 1999). However, reliable and timely information on the nature, extent, spatial distribution pattern and temporal behaviour of mangrove forests (which is prerequisite for restoration and management) is not available. Satellite based remote sensing techniques have proved successful in providing a comprehensive, reliable and up-to-date information on land use, land cover and change dynamics periodically in most cost effective manner. The present paper describes the use of remote sensing to evaluate changes in the mangrove vegetation and other land cover in the Bhitarkanika Sanctuary area from 1988 to 2004. The results of the study would be useful in effective development and management of mangroves.

2. STUDY AREA

Bhitarkanika Wildlife Sanctuary is a rich, lush green vibrant eco-system lying in the estuarine region of Brahmani, Dhamra and Baitarani rivers in the North-Eastern corner of Kendrapara district of Orissa, east coast of India (Fig. 1). It covers an area of 672 sq. km. extending from 86° 48′ E to 87° 03′ E longitudes and 20° 33′ N to 20° 47′ N latitudes. It is surrounded by the Bay of Bengal on the east, the villages of Kendrapara district on the west, Baitarani and the Dhamra rivers on the north and the Mahanadi delta on the south. The area is intersected by a network of creeks with Bay of Bengal on the east. It provides home to well over 215 species of birds including winter migrants from central-Asia and Europe. Giant salt water crocodiles and variety of other wildlife inhabit in this eco-system which forms one of the most spectacular wildlife area of Asia. The Government of Orissa declared this area as a sanctuary in 1975 for better protection of the habitat. Later, the core area (145 sq. km.) of the sanctuary has been declared as a National Park in the year 1998. The total mangrove area is a mixture

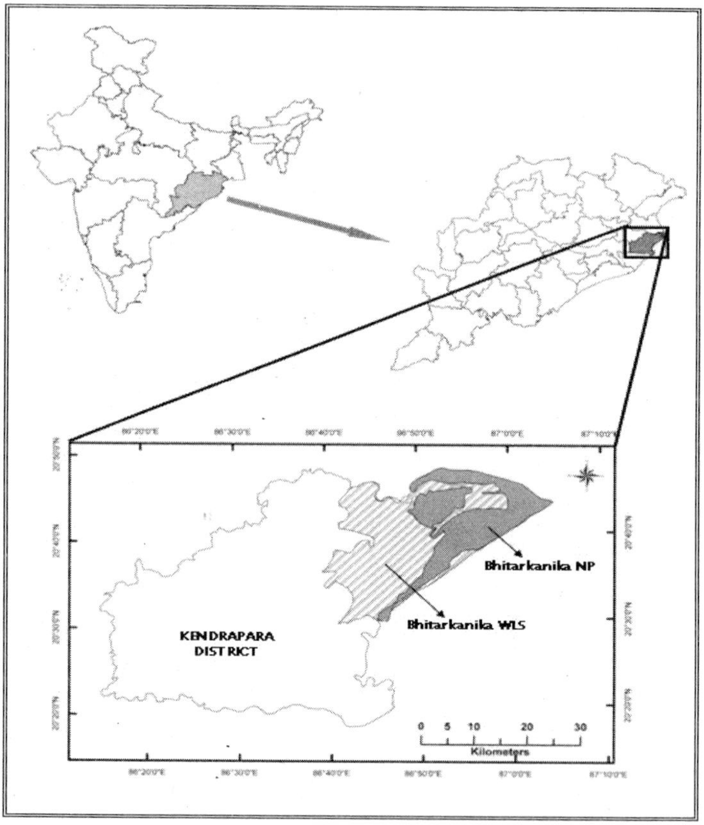

Figure 1: Study area showing Bhitarkanika Wildlife Sanctuary, Orissa.

of 13 protected reserve forests (PRF), 12 protected forests (PF) and one newly formed island (Nayak, 2004). Due to its rich diversity in flora and fauna, this mangrove area has been declared as a Ramsar site, a wetland of international importance in the year 2002.

2.1 Climate and Soil

This area experiences tropical warm and humid climate, with no distinct season. Rain occurs due to south-west monsoon from June to September and the north-east monsoon from December to February. Occasionally these areas experience cyclones. The average rainfall is about 1000 mm, bulk of which is received during June to September. The maximum temperature recorded is 41° C and the minimum is 9° C in the month of May and January respectively. Mean relative humidity ranges from 70% to 95% throughout the year. The soils of the mangrove areas are fine-grained silt or clay formed by the sedimentation of Mahanadi and Brahmani rivers. The soil is frequently inundated with sea water during high tide.

3. MATERIALS AND METHODS

3.1 Data Used

The analysis is based on a multi-temporal satellite imagery study that included the mangroves of Bhitarkanika and the adjacent area, the extent which was analyzed over time. As a result of the analysis, classified thematic maps were prepared to obtain a land-cover map. The satellite data and materials used for the present study are Indian Remote Sensing satellite 1A (IRS-1A) Linear Imaging Self Scanner (LISS-II), IRS P6 (Resourcesat) LISS III and Survey of India (SOI) toposheet no 73 L/10, 73 L/13, 73 L/14, 73 L/15, 73 P/1, 73 P/2 of 1:50,000 scale. All the images were recorded in approximately the same season between December and January, which corresponds to the wet season in the region. All these datasets were geometrically corrected. For geo-referencing, the IRS-P6 LISS III data was co-registered to SOI topographic maps at 1: 50,000 scale using ground control points. The digital data generated was used as a reference for satellite dataset of 1988. All the datasets are in WGS 84 datum and UTM projection. These satellite data were on-screen visually interpreted on a Silicon Graphics workstation using Erdas Imagine version 8.7 image processing software. Ancillary data like SOI toposheets and forest management maps were also used to complement the results of the classification.

3.2 Field Data Collection

Visual interpretation of satellite imagery and reconnaissance survey of the area have been carried out for obtaining patterns of vegetation and other land features during January to December, 2004. Ground truth was collected with the help of False Colour Composite (FCC) hard copy prints of study area 1988 and 2004 (Fig. 2), SOI toposheets, Global Positioning System (GPS) and magnetic compass. The satellite imageries were interpreted and different

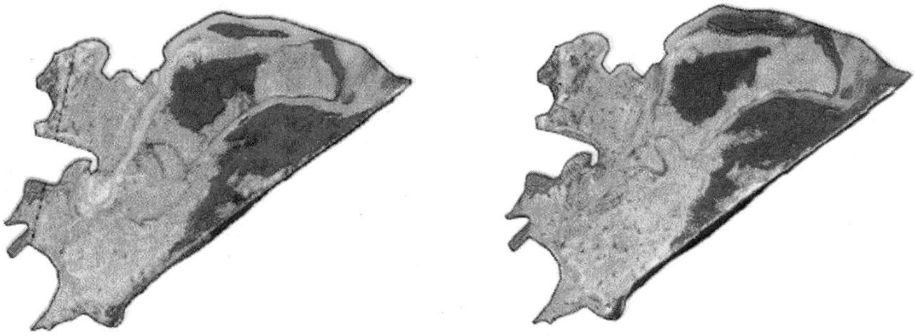

Figure 2: Multitemporal FCC image of Bhitarkanika Wildlife Sanctuary, Orissa (a) IRS 1A LISS II; 17 December, 1988 and (b) IRS P6LISS III; 13 January, 2004.

landuse and landcover categories were delineated on the basis of tone, texture, colour, pattern etc. (Table 1). The field knowledge and interpretation keys have been used for the interpretation and preparation of landuse/landcover maps. The visual interpreted maps of 1988 and 2004 were generated. They were corrected and finalized after thorough ground check. Three maps were overlaid on each other to find out the net change in area of mangroves and other landcover classes between 1988 and 2004 using ARC INFO (9.0) GIS software. The overall distribution of areas in mangrove cover and other land cover categories (1988 and 2004) and the net change in area (1988 to 2004) are estimated (Table 2).

To determine the accuracy of the thematic map obtained using the visual interpretation from the latest 2004 image, an accuracy assessment was carried out. The doubtful areas were identified and the geographic co-ordinates of these points were noted from the visually interpreted classified map. All these points were thoroughly checked in the field with GPS points. The overall accuracy assessment stands at 85%.

Table 1: Image interpretation key for mangrove and other land cover mapping

Land cover class	*Tone*	*Texture*	*Shape*	*Pattern*	*Description*
Mangrove	Dark red	Medium	Varying	Smooth	Tall dense trees
Freshwater swamp	Light red	Medium	Regular	Smooth	Moist and dry deciduous species
Mangrove scrub	Greyish	Coarse	Varying	Rough	Low vegetation density
Littoral scrub	Dark tan	Coarse	Regular	Rough	Scattered vegetation with exposed ground surface
Grasslands	Dark grey	Coarse	Varying	Smooth	Medium size grasses with little herb species
Plantation	Dark brown	Coarse	Regular	Rough	Patchy vegetation along the coastal belt and river beds
Mudflat	Pale blue	Medium	Irregular	Scattered	River sedimentation on the bank
Sand	Whitish	Fine	Regular	Smooth	Mound of sands with sparse vegetation
Water body	Dark or light blue	Smooth	Irregular	Scattered	Rivers and tanks
Agriculture with habitation	Light green or pinkish	Smooth	Regular	Smooth	Crops with current fallow lands

Table 2: Distribution of areas in mangroves and other landuse categories (km^2) in Bhitarkanika Wildlife Sanctuary from 1988-2004

Category	*Area in 1988* (ha)	*Area in 2004* (ha)	*Percentage of total area* 2004	*Change (ha)* (1988-2004)
Mangrove	14639	14718	21.90	79
Mangrove scrub	3100	1658	2.47	–1442
Fresh water swamp	95	95	0.14	0
Littoral scrub	212	41	0.06	–171
Grasslands	25	25	0.04	0
Plantation	18	259	0.39	241
Mudflat	1589	1151	1.71	–438
Sand	437	682	1.01	245
Water body	13310	13758	20.47	448
Agriculture	33777	34815	51.81	1038
Total	67202	67202	100.00	

4. RESULTS AND DISCUSSION

The spatial changes in the mangrove cover have been assessed and the details are given in Tables 2 and 3. The change analysis shows that the major changes took place in the proximity of agricultural lands due to high anthropogenic pressure. Changes were also observed in the river creeks may be due to sedimentation or tidal inundation. Of the total study area, agriculture land with habitation is the major category (51.81%) followed by mangroves

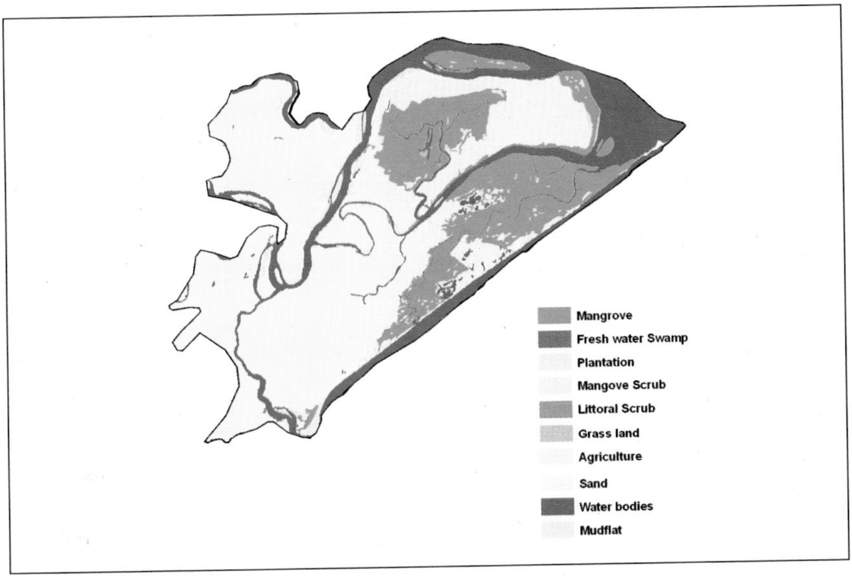

Figure 3: Classified landuse/landcover map of Bhitarkanika Wildlife Sanctuary (2004).

(21.9%), water bodies (20.47%) and mangrove scrub (2.47%) as per 2004 estimate (Table 3). It is observed that there is 79 hectares increase in mangrove cover from 1988 (14,639 hectares) to 2004 (14,718 hectares) as a result of increased protection and consequent regeneration. The overall change statistics of different vegetation cover categories from 1988 to 2004 are presented in Table 3.

The study area represents rich repository of plant wealth. The floristic diversity in each vegetation type and change dynamics in different land cover categories of the study area are shown in Fig. 3 and are discussed below.

4.1 Mangrove Forest

Mangrove forest is typically a closed evergreen forest of moderate height composed of species specially adapted to survive on tidal mud, which is partially submerged with salt water or brackish water. Major area of the National Park is covered by mangrove forest. *Heritiera fomes* occupies major portion of Bhitarkanika Reserve Forest (BRF) followed by *Excoecaria agallocha*, *Avicennia alba* and *Sonneratia apetala*. *Excoecaria agallocha* and *Avicennia alba* occur as a pure community found on the offshore islands or in fringes to the seaward side. *Sonneratia apetala* occurs along the creeks.

It is observed that during 1988-2004, 153 hectares of mangrove forest has been converted to scrub (103 heactares), littoral scrub (41 hectares), water area (6 hectares) and mudflat (3 hectares). On the other hand 194-hectare mangrove forest has been gained from other landuse categories. Interestingly, there is no conversion of mangrove to agriculture due to high protection. The key to their protection lies in the wise management and use of mangrove habitat, and in the enforcement of existing rules and regulations by the State Forest Department.

4.2 Mangrove Scrub

This is an open forest of very low average height, often 3-5 m in height and represents the species of mangrove forest. With increasing pressure of biotic factors, the vegetation is rapidly decreasing. It is characterized by mixed mangroves. Even though, there is no specific zonation pattern, *Excoecaria agallocha* and *Avicennia alba* are the pioneering species found in degraded and scattered mangrove areas. Palm swamp vegetation is also found in drier areas within or outside the mangrove forests mixed with scrub areas on the landward side. Palm swamp shows typical representation of tufted palms (*Phoenix paludosa*) up to 3 m in height. The scrub vegetation facing high anthropogenic pressure and the change in landscape dynamics is clearly noticeable in images. A net change of 1442 hectares is observed due to conversion of scrub area to other landcover categories from 1988 to 2004 (Table 3).

Table 3: Change area matrix from 1988 to 2004

Land cover class	*Mangrove*	*Mangrove scrub*	*Fresh water swamp*	*Littoral scrib*	*Grass-land*	*Planta-tion*	*Mudflat*	*Sand*	*Water body*	*Agricul-ture*	*Total*
Mangrove	14486	103	0	41	0	0	3	0	6	0	14639
Mangrove scrub	31	2526	0	0	0	0	201	0	11	331	3100
Freshwater swamp	0	0	95	0	0	0	0	0	0	0	95
Littoral scrub	0	0	0	0	0	0	0	0	0	0	0
Grassland	0	0	0	0	25	0	0	0	0	0	25
Plantation	0	0	0	0	0	230	0	0	0	0	230
Mudflat	0	21	0	0	0	0	1568	0	0	0	1589
Sand	0	0	0	0	0	0	0	437	0	0	437
Water body	163	0	0	0	0	29	0	0	13118	0	13310
Agriculture	0	0	0	0	0	0	0	0	0	33777	33777
Total	14680	2650	95	41	25	259	1772	437	13135	34108	67202

4.3 Fresh Water (*Diospyros Peregrina* Dominated) Swamp Forest

This type is remarkably pure localized freshwater swamp forest and found above the tide level mainly in Bhitarkanika RF and occupies an area of 95 hectares. The ground is inundated by fresh water. Forest cover is fairly dense and biologically rich with number of plant species. *Diospyros peregrina* is the most dominant tree species in this type, associated with several evergreen tree species and few deciduous elements. There is no remarkable change in the vegetation cover because of its geographic location.

4.4 Littoral (*Tamarix-Salvadora*) Scrub

This type is a representative form of deforested mangrove landscape occurring in semi-arid or arid saline soils covering 41 hectares of the study area. In 1986, it was formed due to mass cutting of mangroves in Sunei-Rupei protected forest (SRPF) area. *Tamarix troupii* and *Salvadora persica* are pioneering woody species found in degraded and deforested mangrove areas and often form pure *Tamarix* and *Tamarix-Salvadora* communities towards the landward zone. It is typically a shrubby formation about 2 to 4 m in height. The vegetation mostly represents exotic herbaceous species like *Parthenium histrophorus*, *Cleome viscosa*, *Croton bonplandianum*, *Hyptis suaveolens*, *Mimosa pudica* and *Scoparia dulcis*.

4.5 Grass Lands

This is a unique type of vegetation found in BRF, nearby fresh water swamp forest admixtures with muddy substratum and occupies an area of 25 hectares. Major portion of the area is covered by tall grasses. *Arundo donax*, *Chrysopogon aciculatus*, *Dicanthium pertusum* and *Imperata cylindrica* are some of the dominant grass species found in this area.

4.6 Mudflats

Mudflat is a wet land in which the salty water inundate the surface of the ground. The vegetation seen in this area is of herbaceous type and generally referred as mudflat (salt marshes) vegetation. Due to high moisture and salinity content, halophytic species like *Suaeda maritima*, *Suaeda nudiflora*, *Suaeda monoica*, *Sesuvium portulacastrum* and *Arthocnemum indicum* grow predominantly. Sanctuary has a significant area under mudflats (1151 hectares), which is main substratum for mangrove regeneration and plantation activities. It has been observed that the mudflat area is gradually decreasing from 1589 hectares in 1988 to 1151 hectares in 2004.

4.7 Sand Dunes

Loose sand mounds along the coastal strip of south eastern side of Bhitarkanika are well covered by the usual runners like *Ipomoea pes-caprae* and *Launaea sarmentosa*. *Spinifex littoreus*, *Hydrophylax maritima* and *Polycarpon prostratum* support the foreshore zone. It covers an area of 682 hectares in the sanctuary.

4.8 Plantation

Orissa Forest Department has raised plantation of *Casuarina equisetifolia* and *Rhizophora mucronata* in the coastal sand and river beds respectively. The plantation area has considerably increased from 230 hectares in 1988 to 259 hectares in 2004 (Table 3). The landcover system undergoes a significant change according to the change in socio-economic and natural conditions of the people. As the people are adopting more to agricultural practice, the agriculture area has increased from 33,777 hectares (1988) to 34,815 hectares (2004). A net change of 1038-hectare area is observed due to conversion of mangrove scrub to agricultural field by the villagers surrounding it. Littoral scrub vegetation was formed due to mangrove deforestation. In course of time, several exotic species, spread over the area, are affecting the native vegetation. The present study has shown a significant increase in mangrove forest area (79 hectares) from 1988 to 2004.

5. CONCLUSIONS

Remote sensing (RS) and Geographical Information System (GIS) are playing a major role in getting a synoptic view of the status of the present vegetation. A comparative study of the present and past conditions of mangrove vegetation brings an overall picture to convince forest officials, managers, decision makers and planners for further conservation and restoration activities. RS in conjunction with GIS technology can play a vital role in the monitoring and planning of the mangrove forest, by multi-temporal interpretation of satellite data. A high spatial resolution data may be able to give more detailed and better information. The information generated for the Bhitarkanika Sanctuary area will aid in understanding the spatial distribution of the mangrove forest and periodical change over more than 16 years which ultimately help the Forest Department, Government of Orissa in further planning and taking appropriate and timely decisions for sustaining the rest of the mangrove cover.

ACKNOWLEDGEMENTS

The authors are grateful to Dr. P.S. Roy, Deputy Director (RS & GIS, Application Area), National Remote Sensing Agency, Hyderabad and Dr. A.K. Biswal, Dept. of Botany, North Orissa University, Baripada for their valuable suggestions and encouragement. We are greatly indebted to Chief Wildlife Warden, Bhubaneswar, Deputy Conservator of Forest, Mangrove forest Division (Wildlife), Rajnagar for granting permission and providing field staffs to do the research work. We thank the Department of Biotechnology (DBT) and Department of Space (DOS) for funding the biodiversity projects under which this study was carried out.

REFERENCES

Bandarnayake, W.M., 1998. Traditional and medicinal uses of mangroves. *Mangroves and Salt Marshes*, **2**: 133-148.

Banerjee, L.K. and Rao, T.A., 1990. Mangroves of Orissa Coast and Their Ecology. Bishen Singh and Mahendrapal Singh (eds.), Dehra Dun.

Banerjee, L.K., 1984. Vegetation of the Bhitarkanika Sanctuary in Cuttack district, Orissa, *India. J. Econ. Tax. Bot.*, **5**: 1065-1079.

Chadha, S.K. and Kar, C.S. (eds), 1999. Bhitarkanika: Myth and Reality. Natraj Publishers, Dehra Dun.

Dahdouh-Guebas, F., Mathenge, C., Kairo, J.G. and Koedam, N., 2000a. Utilization of mangrove wood products around Mida Creek (Kenya) amongst subsistence and commercial users. *Economic Botany*, **54**: 513-527.

Dahdouh-Guebas, F., Verheyden, A., De Genst, W., Hettiarachchi, S. and Koedam, N., 2000b. Four decade vegetation dynamics in Sri Lankan mangroves as detected from sequential aerial photography: a case study in Galle. *B. Mar. Sci.*, **67**: 741-759.

Das, P., Basak, U.C. and Das, A.B., 1997. Restoration of the mangrove vegetation in the Mahanadi delta, Orissa, India. *Mangroves and Salt Marshes*, **1**: 155-161.

Forest Survey of India, 2003. State of Forest Report, Ministry of Environment and Forests, Government of India, New Delhi, pp. 21-22.

Kumar, R., 2000. Distribution of mangroves in Goa. *Indian J. Forestry*, **23**: 360-365.

Ministry of Environment and Forests, Government of India, New Delhi, 1987. Mangroves of India; Status Report, pp. 52-55.

National Remote Sensing Agency, 1983. Mapping of forest cover in India from satellite imagery (1972-75 and 1980-82), Summary Report, Hyderabad, pp. 5-6.

Nayak, A.K., 2004. Pictorial Guide to Mangrove Flora of Bhitarkanika, Mangrove Forest Division (Wildlife), Government of Orissa, Rajnagar, pp. 1-48.

Saxena, H.O. and Brahmam, M., 1996. The Flora of Orissa, Vol I-IV, Orissa Forest Development Corporation, Bhubaneswar.

Spalding, M.D., Blasco, F. and Field, C.D., 1997. World Mangrove Atlas. The International Society for Mangrove Ecosystems, Okinawa, pp. 178-180.

Morphodynamics of Godavari Tidal Inlets

Sarika Jain, P.N. Sridhar, B. Veera Narayan and A. Surendran

National Remote Sensing Agency
Dept of Space, Govt. of India, Balanagar, Hyderabad 500 037

1. INTRODUCTION

Evolution of new features or elope of existing features on the coast may affect the entire coastal system. The morphodynamics of coastal systems implies the nature and time-varying behaviour of coastal landforms and the mechanisms that control this behaviour within specific temporal and spatial scales. Morphodynamic analysis of shoreline helps in understanding the evolution of new feature in response to seasonal and episodic events causing changes in wave direction, alongshore sediment transport, bathymetry and shoreline orientation etc. Changes in inlet morphology depict the mechanism of interaction between inlet currents and long shore sediment transport. Spit is a landform that develops where a re-entrant occurs in bay or river mouth. Geologists consider the development of barrier spit and formation of lagoon environment as a progradation process (Ramakumar, 2000). The movement of sediment along a shore forms spits by long shore drift. Where the direction of the shore turns inland (reenters), the longshore current spreads out or dissipates. No longer are currents able to carry the full load, much of the sediment are dropped. This causes a bar to build out from the shore, eventually becoming a spit. Understanding of spit geometry evolution through analytical models provide several controlling parameters (Nicolas, 1999), which makes the study complicated. This study becomes possible and easy, if temporal remotely sensed data is used. Temporal remote sensing data provides a clear picture of the significant and interesting long term and short tem changes in the spit geometry and evolution. In the east coast of India, rivers are the major sediment supplier to the littoral drift (Chandramohan and Nayak, 1991; Shetye, 1999). The morphodynamics of Godavari river inlets and the Kakinada spit called Hope Island (16° 43′ and 17° 00′ N, and 82° 15′ and 82° 22′ E) is discussed in this paper.

The study area (Fig. 1), Godavari river mouth, receives 38.839×10^9 kg/year sediments as per Central Water Commission (1984) estimate, which plays a significant role in the inlets dynamics and evolution of the Kakinada spit. Among the several tributaries of Godavari River, Nilarevu bifurcates into Coringa and Gaderu forming the second largest mangrove ecosystem of India (Satyanarayana et al., 2001) and neritic waters from Kakinada Bay (Azariah et al., 1992). Mean tidal amplitude of this low meso scale tidal region is around 1.3 m with a semi diurnal variation of 0.72-1.0 m. Significant wave height in the bay is less than 1.2 m (inside) and 1.2-2 m (outside). In the south, at the Nilarevu river mouth significant wave height is 1.8-3 m (Survey of India (SOI) Topography Map, 1975). The estimated southern long shore sediment transport during October to February is around $2.62 \times 10^3 m^3 . y^{-1}$ and the northerly transport during March to September is around $9.60 \times 10^3 m^3 . y^{-1}$, with net drift towards north (Chandramohan et al., 1991).

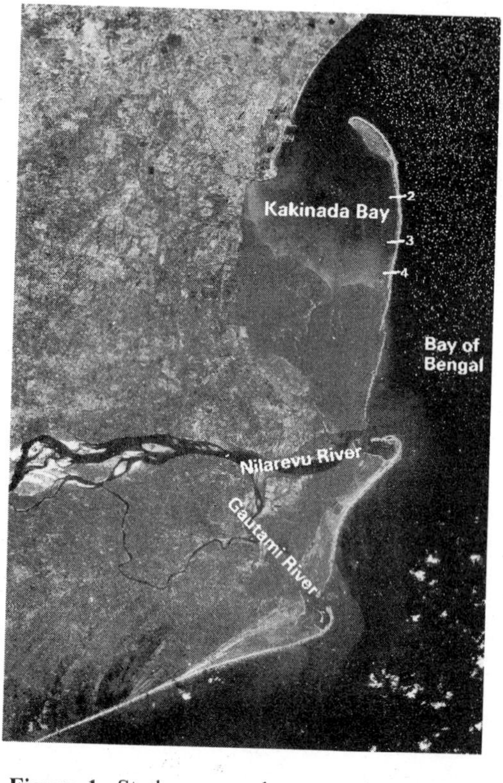

Figure 1: Study area and transect locations.

In the present study, time series multi sensor satellite data for the period 1987, 1997, 2000, 2002 and 2004 with hydrographic and relief data from Naval Hydrographic Office (NHO) and SOI have been used for change detections. Remote sensing data are interpreted to study the long term and short-term changes in the coringa morphology. The preliminary study of temporal satellite data have shown three active morphologic units at the Godavari coast: (1) Elongated stable Kakinada spit trending N-S, (2) Nilarevu river inlet with compound spit trending NE-SW and flood shoal and (3) Gautami river inlet with unstable spit trending south and barrier island. In general, various parameters like length, elongation speed, width, formation of over-wash fans, elevation above mean sea level, depth of closure and the tendency to re-curvature are considered for the study of spit dynamics (Kraus, 1999). Among them spit length, elongation rate and width are considered as the important parameters in the present study of spit dynamics, though other parameters are useful as supporting data.

2. DATA AND METHODS

In this study time series multi spectral satellite data namely Landsat Thematic Mapper (TM) (April, 1987), Indian Remote Sensing Satellite (IRS)-1C- Linear Imaging Self Scanner (LISS-III) (June, 1997) and Enhanced Thematic Mapper (ETM) (December, 2000), IRS-1D LISS III (April, 2002) and IRS-P6 LISS-III (January, 2004), IRS-P4-Ocean Color Monitor (OCM) data (November, 1999 and April, 2000) have been used to study the dynamic changes in the inlet and spit. Baseline information have been extracted from SOI topographic map (1:50,000, 1975) and NHO Chart [chart No. 3009, 1:60,000, 1967] have been used. Satellite data were geocorrected using control points from topographic map. Visual interpretation of satellite data was carried out to create morphological maps using GIS tools by observing the different interpretation elements like tone, texture, shape, association etc. To study the nearshore sediment transport in the study area, suspended sediment concentrations (SSC) have been derived from IRS-P4-OCM data acquired during Nov. 1999 and April 2000. Change detection technique has provided short-term changes during 1987-2004 and long-term changes during 2002-2004. Analyses of spit geometry (spit length, width and elongation rate) at different time intervals make us to understand the relation between inlet dynamics and littoral processes. Three transects were selected to measure the width of the spit from temporal satellite data. Using satellite data of 1987, 1997, 2000, 2002 and 2004 the length of the spit was measured. The shoreline shifting along Kakinada spit was studied with satellite data and bathymetric profiles of 1967 at selected transects. The bathymetric map of Kakinada was retrieved through interpolation technique taking the sea depth points from the NHO chart surveyed during 1967. These depth profiles at selected transects were compared with bathymetric profiles of 1998 given by Tripathi and Rao (2000).

3. RESULTS AND DISCUSSION

The analysis of IRS-P4-OCM derived suspended sediment concentrations (SSC) revealed that the stretch from Gautami Godavari to Kakinada spit behaves as a single cell contributing to the nearshore sediment transport and dynamics with localized sub cells. The trend analysis of Kakinada spit geometry showed sustained long-term unrestricted growth, which is also supported by earlier work on Kakinada spit (Satyaprasad, 1986). According to the study, the spit started evolving since 1848 and the net average growth of the spit is estimated around 123.09 metre/year (m/y) over a period of 139 years. The spit elongation rate estimated from satellite data of 1987 and 2004 over a period of 17 years is approximately 45.29 m/y. The above estimation includes error of about ±11.6% from different sources including image transformation error of ±5.8%. As growth of spit is directly proportional to the long shore sediment transport rate, the unrestricted growth of Kakinada spit shows that there is a constant supply of sediment from south. This growth could sustain until the tidal forcing at the bay channel restricts the spit elongation similar to that of an inlet channel.

The spit analyses show that there is a decline in the spit width during 1987-2000 while elongation rate is negligible during 1987 to 2004 (Fig. 2). In this situation, if net sediment transport within the cell remains constant, when there is no further growth in the length of the spit then the increase in the width of the spit is imminent (Kraus, 1999). Therefore, there was a beach loss in the spit prior or during the study period due to severe erosion.

This is also supported by the clear indication of beach erosion in headland and spit distal portion located adjacent to the Coringa mangroves. As a result of beach erosion, Pillavarava Kaluva channel is directly opened to the sea across the beach in the east (Fig. 3, a: TM 1987; b: IRS-P6-LISS III 2004). Satellite data of IRS-1C LISS III (1997) have shown new dimensions of mangrove areas directly facing towards seawater. Cause for such changes can be viewed as wave erosion and cross-shore sediment drift from distal part of spit during stormy conditions. The years with cyclonic event in the study region are listed here and the year 1996 was such a year with severe cyclonic storm.

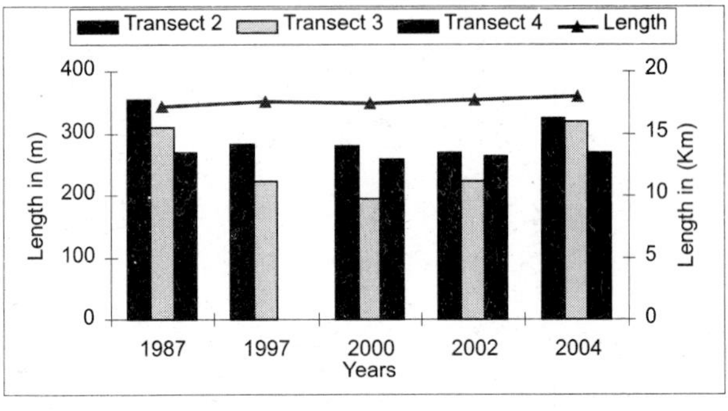

Figure 2: Changes in spit width and length at selected cross-sections in different years.

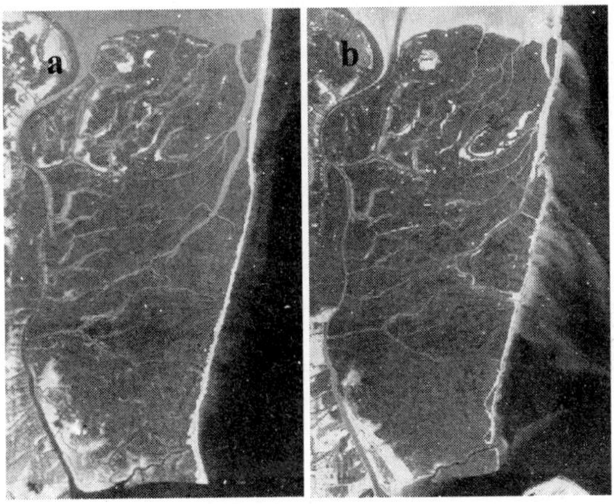

Figure 3: Pillavarava Kalauva channel directly opened to sea.

Years of rough weather and cyclonic events in the study region

1969	1996
1982	1998
1987	2000
1992	

In addition to longshore sediment drift from south, Nilarevu and Gautami rivers play major role in the Kakinada spit growth. Therefore, analyses of inlet morphology help in understanding sediment balance between inlets mouth and Kakinada spit. In this respect, two spits, one at Nilarevu and other at Gautami inlets, gain importance. During our study, compound spit development at the Nilarevu river mouth trending towards the north has been observed (Fig. 4). The Nilarevu spit growth derives the sediment for the

Figure 4: Compound spit formation at Nilarevu river mouth.

south that is from river Gautami-Godavari discharge. The change in the Kakinada spit geometry can be mostly attributed to the sediment from Nilarevu river discharge. The satellite estimated net changes in spit area during 1987-2002 and 2002-2004 at Kakinada, Nilarevu river mouth and Gautami river mouth are given in Table 1. This clearly supports that the change in Kakinada spit geometry is attributed to the lack of sediment from south and erosion due to 1996 cyclone. Reconstruction of the Kakinada spit to the original width from 2000 to 2002 was slow but later on there is an increment in the sediment supply which is evident from the increment in the spit width observed in IRS-P6-LISS III data of 2004.

Table 1: Satellite estimated net changes during 1987-2002 and 2002-2004

Area	*Period*	*Erosion (sq. km.)*	*Deposition (sq. km.)*	*Net Change (sq. km.)*	*Rate of Change (sq. km./y)*
Kakinada	1987-2002	3.063	0.521	-2.542	0.169
	2002-2004	0.687	1.089	0.402	0.201
Nilarevu	1987-2002	2.058	3.20	1.142	0.076
	2002-2004	0.925	0.783	-0.142	-0.071
Gautami Godavari	1987-2002	2.125	1.24	-0.885	-0.059
	2002-2004	1.801	1.309	-0.492	-0.246

The presence of spit at the southern tip of Gautami-Godavari mouth observed from TM data of 1987 and ETM data of 2000 and accumulation of sediment carried by the river discharge caused the bypass of inlet channel to the north (Fig. 5). However, the formation of the spit at the northern tip of the mouth is found to be temporal as seen in the TM of 1987, ETM of 2000 and IRS-P6-LISS III of 2004 data. The field verification shows that spit growth is temporary in nature depending upon the sediment supply from southerly drift and river discharge rate during monsoon.

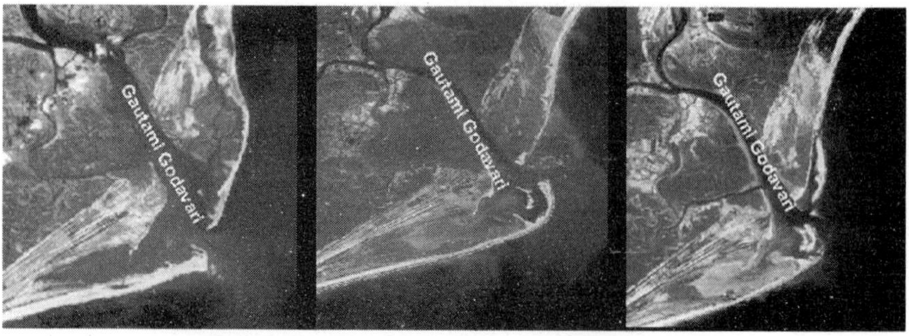

Figure 5: Changes at Gautami-Godavari river mouth.

4. CONCLUSIONS

The elongation speed of Kakinada spit during the study period is uniform and temporal change in the spit width at different transects has been observed from satellite data. The average spit growth has been retarded and landward shifting has been observed during this period due to deficiency in the supply of sediment from the south and lack of nourishment to rebuild the beach eroded during 1996 cyclone. Over a period of 17 years short-term dynamics of spit indicated a net balance in sediment transport within the cell depending upon the local conditions like river discharge, alongshore transport, stability of depositional feature in the updrift direction and episodic events like storms.

REFERENCES

Azariah, J., Azariah, H., Gunasekaran, S. and Selvam, V., 1992. Structure and species distribution in Coringa mangrove forest, Godavari Delta, Andhra Pradesh, India. *Hydrobiologia*, **247:** 11-16.

Chandramohan, P. and Nayak, B.U., 1991. Long shore sediment transport along the Indian coast. *Indian Journal of Marine Sciences*, **20:** 110-114.

Nicholas, C.D., 1999. Analytical model of spit evolution at inlets. *Proc. Coastal Sediments* 99, ASCE, pp. 1739-1754.

Ramakumar, M., 2000. Recent Changes in Kakinada Spit, Godavari Delta. *Journal Geological Society of India*, **55:** 183-188.

Satyanarayana, B., Raman, A.V., Dehairs, F., Kalavati, C. and Chandramohan, P., 2001. Mangrove floristic and zonation patterns of Coringa, Kakinada Bay, East Coast of India. *Wetlands Ecology and Management*, **10:** 25-39.

Satyaprasad, D., 1986. Morphodynamics of the beaches and sand spit, Kakinada Bay, East Coast of India, Andhra Pradesh. Andhra University, PhD Thesis.

Shetye, S.R., 1999. Dynamics of circulation of the water around India. *In:* Somayajulu, B.L.K (ed), Ocean Science: Trends and Future Directions. Akademia Books International, New Delhi, India, pp.1-21.

Tripathi, N.K. and Rao, A.M., 2001. Investigation of erosion on Hope Island using IRS-1D LISS-III data. *International Journal of Remote Sensing*, **22(5):** 883-888.

Index